1992

Thinking

Recursively

Thinking Recursively

ERIC ROBERTS
Department of Computer Science
Wellesley College, Wellesley, Mass.

JOHN WILEY & SONS, INC.
New York · Chichester · Brisbane · Toronto · Singapore

Library of Congress Cataloging in Publication Data:

Roberts, Eric.
   Thinking recursively.

   Bibliography: p.
   Includes index.
   1. Recursion theory.   I. Title.
QA9.6.R63   1985   5113   85-20365
ISBN 0-471-81652-3

Printed in the United States of America

10 9 8 7 6 5

In loving memory of my grandmother
RUTH STENIUS ROBERTS
(1899–1982)

Preface

In my experience, teaching students to use recursion has always been a difficult task. When it is first presented, students often react with a certain suspicion to the entire idea, as if they had just been exposed to some conjurer's trick rather than a new programming methodology. Given that reaction, many students never learn to apply recursive techniques and proceed to more advanced courses unable to write programs which depend on the use of recursive strategies. This book is intended to demystify this material and encourage the student to "think recursively."

This book is intended for use as a supplementary text in an intermediate course in data structures, but it could equally well be used with many other courses at this level. The only prerequisite for using this text is an introductory programming course. Since Pascal is used in the programming examples, the student must also become familiar with Pascal programming, although this can easily be included as part of the same course. To support the concurrent presentation of Pascal and the material on recursion, the programming examples in the early chapters require only the most basic features of Pascal.

In order to develop a more complete understanding of the topic, it is important for the student to examine recursion from several different perspectives. Chapter 1 provides an informal overview which examines the use of recursion outside the context of programming. Chapter 2 examines the underlying mathematical concepts and helps the student develop an appropriate conceptual model. In particular, this chapter covers mathematical induction and computational complexity in considerable detail. This discussion is designed to be nonthreatening to the math-anxious student and, at the same time, include enough formal structure to emphasize the extent to which computer science depends on mathematics for its theoretical foundations.

Chapter 3 applies the technique of recursive decomposition to various mathematical functions and begins to show how recursion is represented in Pascal. Chapter 4 continues this discussion in the context of recursive procedures, emphasizing the parallel between recursive decomposition and the more familiar technique of stepwise refinement.

Chapters 5 through 9 present several examples of the use of recursion to solve increasingly sophisticated problems. Chapter 7 is of particular importance and covers recursive sorting techniques, illustrating the applicability of the recursion methodology to practical problem domains. Chapter 9 contains many delightful examples, which make excellent exercises and demonstrations if graphical hardware is available.

Chapter 10 examines the use of recursive procedures in the context of recursive data structures and contains several important examples. Structurally, this chapter appears late in the text primarily to avoid introducing pointers prematurely. For courses in which the students have been introduced to pointers relatively early, it may be useful to cover the material in Chapter 10 immediately after Chapter 7.

Finally, Chapter 11 examines the underlying implementation of recursion and provides the final link in removing the mystery. On the other hand, this material is not essential to the presentation and may interfere with the student's conceptual understanding if presented too early.

I am deeply grateful for the assistance of many people who have helped to shape the final form of the text. I want to express a special note of appreciation to Jennifer Friedman, whose advice has been invaluable in matters of both substance and style. I would also like to thank my colleagues on the Wellesley faculty, Douglas Long, K. Wojtek Przytula, Eleanor Lonske, James Finn, Randy Shull, and Don Wolitzer for their support. Joe Buhler at Reed College, Richard Pattis at the University of Washington, Gary Ford at the University of Colorado at Colorado Springs, Steve Berlin at M.I.T., and Suzanne Rodday (Wellesley '85) have all made important suggestions that have dramatically improved the final result.

Eric Roberts

Contents

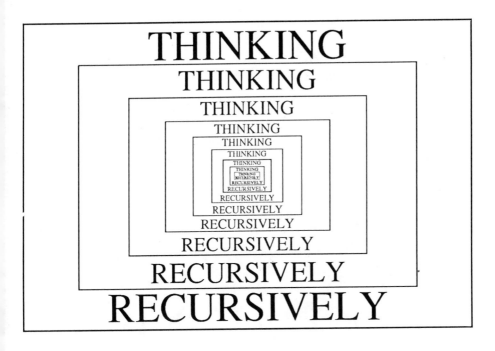

The Idea of Recursion

Of all ideas I have introduced to children, recursion stands out as the
one idea that is particularly able to evoke an excited response.
—Seymour Papert, Mindstorms

At its essence, computer science is the study of problems and their solutions.
More specifically, computer science is concerned with finding systematic procedures that guarantee a correct solution to a given problem. Such procedures
are called *algorithms*.

This book is about a particular class of algorithms, called *recursive algorithms*, which turn out to be quite important in computer science. For many
problems, the use of recursion makes it possible to solve complex problems
using programs that are surprisingly concise, easily understood, and algorithmically efficient. For the student seeing this material for the first time, however,
recursion appears to be obscure, difficult, and mystical. Unlike other problem-
solving techniques which have closely related counterparts in everyday life,
recursion is an unfamiliar idea and often requires thinking about problems in
a new and different way. This book is designed to provide the conceptual tools
necessary to approach problems from this recursive point of view.

Informally, *recursion* is the process of solving a large problem by reducing
it to one or more *subproblems* which are (1) identical in structure to the original
problem and (2) somewhat simpler to solve. Once that original subdivision has
been made, the same decompositional technique is used to divide each of these
subproblems into new ones which are even less complex. Eventually, the subproblems become so simple that they can be then solved without further subdivision, and the complete solution is obtained by reassembling the solved
components.

1-1 An Illustration of the Recursive Approach

Imagine that you have recently accepted the position of funding coordinator
for a local election campaign and must raise $1000 from the party faithful. In
this age of political action committees and direct mail appeals, the easiest

approach is to find a single donor who will contribute the entire amount. On the other hand, the senior campaign strategists (fearing that this might be interpreted as a lack of commitment to democratic values) have insisted that the entire amount be raised in contributions of exactly $1. How would you proceed?

Certainly, one solution to this problem is to go out into the community, find 1000 supporters, and solicit $1 from each. In programming terms, such a solution has the following general structure.

```
PROCEDURE COLLECT1000;
   BEGIN FOR I := 1 TO 1000 DO
   Collect one dollar from person I
   END;
```

Since this is based on an explicit loop construction, it is called an *iterative* solution.

Assuming that you can find a thousand people entirely on your own, this solution would be effective, but it is not likely to be easy. The entire process would be considerably less exhausting if it were possible to divide the task into smaller components, which can then be delegated to other volunteers. For example, you might enlist the aid of ten people and charge each of them with the task of raising $100. From the perspective of each volunteer, the new problem has exactly the same form as the original task. The only thing which has changed is the dimension of the problem. Instead of collecting $1000, each volunteer must now collect only $100—presumably a simpler task.

The essence of the recursive approach lies in applying this same decomposition repeatedly at each stage of the solution. Thus, each volunteer who must collect $100 finds ten people who will raise $10 each. Each of these, in turn, finds ten others who agree to raise $1. At this point, however, we can adopt a new strategy. Since the campaign can accept $1 contributions, the problem need not be subdivided further into dimes and pennies, and the volunteer can simply contribute the necessary dollar. In the parlance of recursion, $1 represents a *simple case* for the fund-raising problem, which means that it can be solved directly without further decomposition.

Solutions which operate in this way are often referred to as "divide-and-conquer" strategies, since they depend on splitting a problem into more manageable components. The original problem divides to form several simpler subproblems, which, in turn, "branch" into a set of simpler ones, and so on, until the simple cases are reached. If we represent this process diagrammatically, we obtain a *solution tree* for the problem:

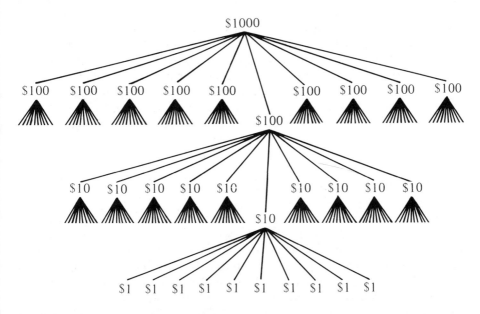

In order to represent this algorithm in a form more suggestive of a programming language, it is important to notice that there are several different *instances* of a remarkably similar problem. In the specific case shown here, we have the independent tasks "collect $1000", "collect $100", "collect $10" and "collect $1", corresponding to the different levels of the hierarchy. Although we could represent each of these as a separate procedure, such an approach would fail to take advantage of the structural similarity of each problem. To exploit that similarity, we must first generalize the problem to the task of collecting, not some specific amount, but an undetermined sum of money, represented by the symbol N.

The task of collecting N dollars can then be broken down into two cases. First, if N is $1, we simply contribute the money ourselves. Alternatively, we find ten volunteers and assign each the task of collecting one-tenth the total revenue. This structure is illustrated by the procedure skeleton shown below:

```
PROCEDURE COLLECT(N);
  BEGIN
    IF N is $1 THEN
      Contribute the dollar directly
    ELSE
      BEGIN
        Find 10 people;
        Have each collect N/10 dollars;
        Return the money to your superior
      END
  END;
```

The structure of this "program" is typical of recursive algorithms represented in a programming language. The first step in a recursive procedure consists of a test to determine whether or not the current problem represents a simple case. If it does, the procedure handles the solution directly. If not, the problem is divided into subproblems, each of which is solved by applying the same recursive strategy. In this book, we will consider many recursive programs that solve problems which are considerably more detailed. Nonetheless, all of them will share this underlying structure.

1-2   Mondrian and Computer Art

During the years 1907 to 1914, a new phase of the modern art movement flourished in Paris. Given the name Cubism by its detractors, the movement was based on the theory that nature should be represented in terms of its primitive geometrical components, such as cylinders, cones, and spheres. Although the Cubist community was dissolved by the outbreak of World War I, the ideas of the movement remained a powerful force in shaping the later development of abstract art. In particular, Cubism strongly influenced the work of the Dutch painter Piet Mondrian, whose work is characterized by rigidly geometrical patterns of horizontal and vertical lines.

The tendency in Mondrian's work toward simple geometrical structure makes it particularly appropriate for computer simulation. Many of the early attempts to generate "computer art" were based on this style. Consider, for example, the following abstract design:

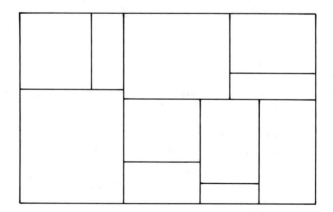

In this example, the design consists of a large rectangle broken up into smaller rectangles by a sequence of horizontal and vertical lines.

For now, we will limit our concern to finding a general strategy for generating a design of this sort and will defer the details of the actual program until Chapter 9, where this example is included as an exercise. To discover this strategy and understand how recursion is involved, it helps to go back to the beginning and follow this design through its evolutionary history. As with any work of art (however loosely the term is applied in this case), our design started as an empty rectangular "canvas":

The first step in the process was to divide this canvas into two smaller rectangles with a single vertical line. From the finished drawing, we can see that there is only one line which cuts across the entire canvas. Thus, at some point early on in the history of this drawing, it must have appeared as follows:

But now what? From here, the simplest way to proceed is to consider each of the two remaining rectangles as a new empty canvas, admittedly somewhat smaller in size. Thus, as part of the process of generating a "large" Mondrian drawing, we have reduced our task to that of generating two "medium-sized" drawings, which, at least insofar as the recursive process is concerned, is a somewhat simpler task.

As a practical matter, we must choose one of these two subproblems first and work on it before returning to the other. Here, for instance, we might choose to work on the left-hand "subcanvas" first and, when we finish with that, return to finish the right-hand one. For the moment, however, we can forget about the right-hand part entirely and focus our attention on the left-hand side. Conceptually, this is a new problem of precisely the original form. The only difference is that our new canvas is smaller in size.

Once again, we start by dividing this into two subproblems. Here, since the figure is taller than it is wide, a horizontal division seems more appropriate, which gives us:

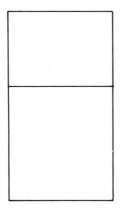

Just as before, we take one of these smaller figures and leave the other aside for later. Note that, as we proceed, we continue to set aside subproblems for future solution. Accumulating a list of unfinished tasks is characteristic of recursive processes and requires a certain amount of bookkeeping to ensure that all of these tasks do get done at some point in the process. Ordinarily, the programmer need not worry about this bookkeeping explicitly, since it is performed automatically by the program. The details of this process are discussed in Chapter 5 and again in Chapter 11.

Eventually, however, as the rectangles become smaller and smaller, we reach a point at which our aesthetic sense indicates that no further subdivision is required. This constitutes the "simple case" for the algorithm—when a rectangle drops below a certain size, we are finished with that subproblem and must return to take care of any uncompleted work. When this occurs, we simply consult our list of unfinished tasks and return to the one we most recently set aside, picking up exactly where we left off. Assuming that our recursive solution operates correctly, we will eventually complete the entire list of unfinished tasks and obtain the final solution.

1-3   Characteristics of Recursive Algorithms

In each of the examples given above, finding simpler subproblems within the context of a larger problem was a reasonably easy task. These problems are naturally suited to the divide-and-conquer strategy, making recursive solutions particularly appropriate.

In most cases, the decision to use recursion is suggested by the nature of the problem itself.* To be an appropriate candidate for recursive solution, a problem must have three distinct properties:

*At the same time, it is important to recognize that "recursiveness" is a property of the *solution* to a problem and not an attribute of the problem itself. In many cases, we can take a problem which seems recursive in its structure and choose to employ an iterative solution. Similarly, recursive techniques can be used to solve problems for which iteration appears more suitable.

1. It must be possible to decompose the original problem into simpler instances of the same problem.
2. Once each of these simpler subproblems has been solved, it must be possible to combine these solutions to produce a solution to the original problem.
3. As the large problem is broken down into successively less complex ones, those subproblems must eventually become so simple that they can be solved without further subdivision.

For a problem with these characteristics, the recursive solution follows in a reasonably straightforward way. The first step consists of checking to see if the problem fits into the "simple case" category. If it does, the problem is solved directly. If not, the entire problem is broken down into new subsidiary problems, each of which is solved by a recursive application of the algorithm. Finally, each of these solutions is then reassembled to form the solution to the original problem.

Representing this structure in a Pascal-like form gives rise to the following *template* for recursive programs:

```
PROCEDURE SOLVE(instance);
  BEGIN
    IF instance is easy THEN
      Solve problem directly
    ELSE
      BEGIN
        Break this into new instances I1, I2, etc.;
        SOLVE(I1); SOLVE(I2); . . . and so forth . . .;
        Reassemble the solutions
      END
  END;
```

1-4   Nonterminating Recursion

In practice, the process of ensuring that a particular decomposition of a problem will eventually result in the appropriate simple cases requires a certain amount of care. If this is not done correctly, recursive processes may get locked into cycles in which the simple cases are never reached. When this occurs, a recursive algorithm fails to *terminate,* and a program which is written in this way will continue to run until it exhausts the available resources of the computer.

For example, suppose that the campaign fund raiser had adopted an even lazier attitude and decided to collect the $1000 using the following strategy:

Find a single volunteer who will collect $1000.

If this volunteer adopts the same strategy, and every other volunteer follows in like fashion, the process will continue until we exhaust the available pool of volunteers, even though no money at all will be raised. A more fanciful example of this type of failure is shown in the following song.

THERE'S A HOLE IN THE BUCKET

Traditional

There's a hole in the bucket, dear Liza, dear Liza
There's a hole in the bucket, dear Liza, a hole
Then fix it, dear Charlie, dear Charlie
Then fix it, dear Charlie, dear Charlie, fix it

With what shall I fix it, dear Liza, dear Liza
With a straw, dear Charlie, dear Charlie

But the straw is too long, dear Liza, dear Liza
Then cut it, dear Charlie, dear Charlie

With what shall I cut it, dear Liza, dear Liza
With a knife, dear Charlie, dear Charlie

But the knife is too dull, dear Liza, dear Liza
Then sharpen it, dear Charlie, dear Charlie

With what shall I sharpen it, dear Liza, dear Liza
With a stone, dear Charlie, dear Charlie

But the stone is too dry, dear Liza, dear Liza
Then wet it, dear Charlie, dear Charlie

With what shall I wet it, dear Liza, dear Liza
With water, dear Charlie, dear Charlie

But how shall I fetch it, dear Liza, dear Liza
In a bucket, dear Charlie, dear Charlie

There's a hole in the bucket, dear Liza, dear Liza,
There's a hole in the bucket, dear Liza, a hole

1-5   Thinking about Recursion—Two Perspectives

The principal advantage of recursion as a solution technique is that it provides an excellent mechanism for managing complexity. No matter how difficult a problem at first appears, if we can determine a way to break that problem down into simpler problems of the same form, we can define a strategy for producing a complete solution. As programmers, all we need to specify is (1) how to simplify a problem by recursive subdivision, (2) how to solve the simple cases, and (3) how to reassemble the partial solutions.

For someone who is just learning about recursion, it is very hard to believe that this general strategy is powerful enough to solve a complex problem. Given a particular problem, it is tempting to insist on seeing the solution in all its gory detail. Unfortunately, this has the effect of reintroducing all the complexity that the recursive definition was designed to conceal. By giving in to skepticism, the usual result is that one takes a hard problem made simpler through recursion and proceeds to make it difficult again. Clearly, this is not the optimal approach and requires finding a new way to think about recursion.

The difference between the perspective of the programmer with considerable experience in recursion and that of the novice is perhaps best defined in terms of the philosophical contrast between "holism" and "reductionism." In *Godel, Escher, Bach*, Douglas Hofstadter defines these concepts by means of the following dialogue:

Achilles   I will be glad to indulge both of you, if you will first oblige me, by telling me the meaning of these strange expressions, "holism" and "reductionism."

Crab   Holism is the most natural thing in the world to grasp. It's simply the belief that "the whole is greater than the sum of its parts." No one in his right mind could reject holism.

Anteater   Reductionism is the most natural thing in the world to grasp. It's simply the belief that "a whole can be understood completely if you understand its parts, and the nature of their 'sum'." No one in her left brain could reject reductionism.

Even though recursion acts as a reductionistic process in the sense that each problem is reduced to a sum of its parts, writing recursive programs tends to require a holistic view of the process. It is the big picture which is important, not the details. In developing a "recursive instinct," one must learn to stop analyzing the process after the first decomposition. The rest of the problem will take care of itself, and the details tend only to confuse the issue. When one cannot see the forest for the trees, it is of very little use to examine the branches, twigs, and leaves.

For beginners, however, this holistic perspective is usually difficult to maintain. The temptation to look at each level of the process is quite strong, particularly when there is doubt about the correctness of the algorithm. Overcoming that temptation requires considerable confidence in the general mechanism of recursion, and the novice has little basis for that confidence.

Achieving the necessary confidence often requires the programmer to adopt a strategy called the "recursive leap of faith." As a strategy, this means that one is allowed to assume the solution to simpler problems when trying to solve a complex one. At first, this appears mystical to the point of being suspect. By becoming more familiar with recursion and by understanding its theoretical basis, however, this faith will be justified and lose its mystical quality.

Even so, it is probably impossible to avoid completely the tendency to undertake the reductionistic analysis. Seeing the details has the effect of justifying the faith required to adopt the more holistic perspective, since one can actually "see it work." Seemingly, this is one of those techniques which must be learned independently, with experience as the best teacher.

Glinda   You've always had the power to go back to Kansas.

Scarecrow   Then why didn't you tell her before?

Glinda   Because she wouldn't have believed me. She had to learn it for herself.

—The Wizard of Oz

Bibliographic Notes

The general nature and philosophy of recursion form the basis of Douglas Hofstadter's Pulitzer Prize winning book *Godel, Escher, Bach* [1979], which includes an important discussion of the distinction between the holistic and reductionistic view. The use of recursion as a general problem-solving tool is also discussed briefly in Wickelgren [1974].

The example of the "money tree" used in this chapter is the invention of Wayne Harvey at the Stanford Research Institute and was conveyed to me by Richard Pattis of the University of Washington. Further information concerning the use of recursive algorithms in the design and construction of "computer art" may be found in Levitt [1976].

Exercises

1-1. Given the structure of the "collect N dollars" algorithm suggested by the **COLLECT** procedure on page 1, what would happen if the original target were 500 instead of 1000? How could this problem be fixed?

1-2. Using a large sheet of graph paper and a note pad to keep track of your unfinished tasks, follow the steps required to generate a Mondrian rectangle such as the one illustrated in the text. Use your own sense of aesthetics to determine where to divide each rectangle and to decide whether or not a rectangle should be divided at all.

1-3. Suppose that there is a pile of sixteen coins on a table, one of which is a counterfeit weighing slightly less than the others. You also have a two-pan balance which allows you to weigh one set of coins against another. Using the divide-and-conquer strategy, how could you determine the counterfeit coin in four weighings?

 If you solve this problem, see if you can come up with a procedure to find the counterfeit coin in just *three* weighings. The strategy is much the same, but the problem must be subdivided in a different way. Can you generalize this approach so that it works for any set of N coins?

Mathematical Preliminaries

One cannot escape the feeling that these mathematical formulae have an independent existence and an intelligence of their own, that they are wiser than we are, wiser even than their discoverers.—*Heinrich Hertz*

For many students who have become interested in computer science through their experience with programming, the notion that computer science requires a strong mathematical foundation is met with a certain level of distrust or disapproval. For those students, mathematics and programming often seem to represent antithetical aspects of the science of computing; programming, after all, is usually fun, and mathematics is, to many, quite the opposite.

In many cases, however, understanding the mathematical foundations of computer science can provide practical insights which dramatically affect the programming process. Recursive programming is an important case in point. As outlined in Chapter 1, programmers often find the concept of recursion difficult primarily because they lack faith in its correctness. In attempting to supply this faith, experience is critically important. For this reason, much of this book consists of programming examples and related exercises which reinforce the skills and tactics required for recursive programming. On the other hand, the ability to prove, through a reasonably cogent mathematical argument, that a particular recursive algorithm does what it is supposed to do also serves to increase one's level of confidence in the underlying process.

This chapter addresses two separate mathematical issues that arise in any complete discussion of recursion. The first is *mathematical induction,* which provides a powerful tool for proving the correctness of certain useful formulae. The second issue is that of *computational complexity,* which is considered here in a very elementary form. Throughout the text, it will often be necessary to compare the efficiency of several different algorithms for solving a particular problem. Determining the computational complexity of each algorithm will provide a useful standard for making that comparison.

2-1   Mathematical Induction

Recursive thinking has a parallel in mathematics which is called *mathematical induction*. In both techniques, one must (1) determine a set of *simple cases* for which the proof or calculation is easily handled and (2) find an appropriate *rule* which can be repeatedly applied until the complete solution is obtained. In recursive applications, this process begins with the complex cases, and the rule successively reduces the complexity of the problem until only simple cases are left. When using induction, we tend to think of this process in the opposite direction. We start by proving the simple cases, and then use the inductive rule to derive increasingly complex results.

The nature of an inductive proof is most easily illustrated in the context of an example. In many mathematical and computer science applications, we need to compute the sum of the integers from 1 up to some maximum value N. We could certainly calculate this number by taking each number in order and adding it to a running total. Unfortunately, calculating the sum of the integers from 1 to 1000 by this method would require 1000 additions, and we would quickly tire of performing this calculation in longhand. A much easier approach involves using the mathematical formula

$$1 + 2 + 3 + \cdots + N = \frac{N(N+1)}{2}$$

This is certainly more convenient, but only if we can be assured of its correctness.

For many formulae of this sort, mathematical induction provides an ideal mechanism for proof. In general, induction is applicable whenever we are trying to prove that some formula is true for every positive number N.*

The first step in an inductive proof consists of establishing that the formula holds when N = 1. This constitutes the *base step* of an inductive proof and is quite easy in this example. Substituting 1 for N in the right-hand side of the formula and simplifying the result gives

$$\frac{1(1+1)}{2} = \frac{2}{2} = 1 \quad .$$

The remaining steps in a proof by induction proceed as follows:

1. Assume that the formula is true for some arbitrary number N. This assumption is called the *inductive hypothesis*.
2. Using that hypothesis, establish that the formula holds for the number N+1.

*Mathematically speaking, induction is conventionally used to prove that some *property* holds for any positive integer N, rather than to establish the correctness of a formula. In this chapter, however, each of the examples is indeed a formula, and it is clearer to define induction in that domain.

Thus, in our current example, the inductive hypothesis consists of making the assumption that

$$1 + 2 + 3 + \cdots + N = \frac{N(N+1)}{2}$$

holds for some unspecified number N. To complete the induction proof, it is necessary to establish that

$$1 + 2 + 3 + \cdots + (N+1) = \frac{(N+1)(N+2)}{2}$$

Look at the left-hand side of the expression. If we fill in the last term represented by the ellipsis (i.e., the term immediately prior to $N+1$), we get

$$1 + 2 + 3 + \cdots + N + (N+1)$$

The first N terms in that sum should look somewhat familiar, since they are precisely the left-hand side of the inductive hypothesis. The key to inductive proofs is that we are allowed to use the result for N during the derivation of the $N+1$ case. Thus, we can substitute in the earlier formula and complete the derivation by simple algebra:

$$\underbrace{1 + 2 + 3 + \cdots + N}_{} + (N+1)$$

$$\frac{N(N+1)}{2} + (N+1)$$

$$= \frac{N^2 + N}{2} + \frac{2N + 2}{2}$$

$$= \frac{N^2 + 3N + 2}{2}$$

$$= \frac{(N+1)(N+2)}{2}$$

Even though mathematical induction provides a useful mechanism for proving the correctness of this formula, it does not offer much insight into how such a formula might be derived. This intuition often has its roots in the geometry of the structure. In this example, the successive integers can be represented as lines of dots arranged to form a triangle as shown:

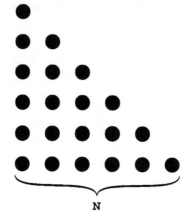

Clearly, the sum of the first N integers is simply the number of dots in the triangle. As of yet, we have not simplified the problem but have merely changed its form. To determine the number of dots, we must apply some geometrical insight. If, for example, we take an identical triangle, invert it, and write it above the first triangle, we get a rectangular array of dots which has exactly twice as many dots as in the original triangle:

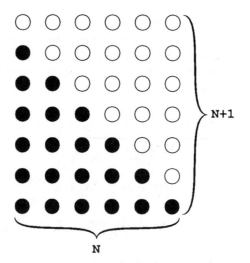

Fortunately, counting the number of dots in a rectangle is a considerably easier task, since the number of dots is simply the number of columns times the number of rows. Since there are N columns and N + 1 rows in this diagram, there are N $\times$ (N + 1) dots in the rectangle. This gives

$$\frac{N(N+1)}{2}$$

as the number of dots in the original triangle.

As a second example, suppose we want to find a simple formula for computing the sum of the first N odd integers:

$$1 + 3 + 5 + \cdots + (\text{Nth odd number})$$

The expression "(Nth odd number)" is a little cumbersome for mathematical manipulation and can be represented more concisely as "$2N-1$".

$$1 + 3 + 5 + \cdots + (2N-1)$$

Once again, we can gain some insight by considering a geometric representation. If we start with a single dot and add three dots to it, those three dots can be arranged to form an L shape around the original, creating a 2×2 square. Similarly, if we add five more dots to create a new row and column, we get a 3×3 square. Continuing with this pattern results in the following figure:

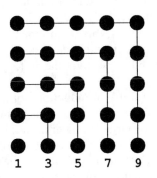

Since we have both an extra row and column, each new L-shaped addition to the figure requires two more dots than the previous one. Given that we started with a single dot, this process therefore corresponds to adding the next odd number each time. Using this insight, the correct formula for the sum of the first N odd numbers is simply

$$1 + 3 + 5 + \cdots + (2N-1) = N^2$$

Although the geometric argument above can be turned into a mathematically rigorous proof, it is simpler to establish the correctness of this formula by induction. One such proof follows. Both the base step and the inductive derivation are reasonably straightforward. Give it a try yourself before turning the page.

Inductive proof for the formula

$$1 + 3 + 5 + \cdots + (2N - 1) = N^2$$

Base step: $1 = 1^2$

Inductive derivation:

$$\underbrace{1 + 3 + 5 + \cdots + (2N - 1)} + 2(N + 1) - 1$$
$$N^2 + 2(N + 1) - 1$$
$$= N^2 + 2N + 1$$
$$= (N + 1)^2$$

There are several ways to visualize the process of induction. One which is particularly compelling is to liken the process of an inductive proof to a chain of dominos which are lined up so that when one is knocked over, each of the others will follow in sequence. In order to establish that the entire chain will fall under a given set of circumstances, two things are necessary. To start with, someone has to physically knock over the first domino. This corresponds to the base step of the inductive argument. In addition, we must also know that, whenever any domino falls over, it will knock over the next domino in the chain. If we number the dominos, this requirement can be expressed by saying that whenever domino N falls, it must successfully upset domino $N + 1$. This corresponds to using the inductive hypothesis to establish the result for the next value of N.

More formally, we can think of induction not as a single proof, but as an arbitrarily large sequence of proofs of a similar form. For the case $N = 1$, the proof is given explicitly. For larger numbers, the inductive phase of the proof provides a mechanism to construct a complete proof for any larger value. For example, to prove that a particular formula is true for $N = 5$, we could, in principal, start with the explicit proof for $N = 1$ and then proceed as follows:

Since it is true for $N = 1$, I can prove it for $N = 2$.
Since I know it is true for $N = 2$, I can prove it for $N = 3$.
Since I know it is true for $N = 3$, I can prove it for $N = 4$.
Since I know it is true for $N = 4$, I can prove it for $N = 5$.

In practice, of course, we are not called upon to demonstrate a complete proof for any value, since the inductive mechanism makes it clear that such a derivation would be possible, no matter how large a value of N is chosen.

Recursive algorithms proceed in a very similar way. Suppose that we have a problem based on a numerical value for which we know the answer when $N = 1$. From there, all that we need is some mechanism for calculating the result for any value N in terms of the result for $N - 1$. Thus, to compute the solution when $N = 5$, we simply invert the process of the inductive derivation:

To compute the value when N = 5, I need the value when N = 4.
To compute the value when N = 4, I need the value when N = 3.
To compute the value when N = 3, I need the value when N = 2.
To compute the value when N = 2, I need the value when N = 1.
I know the value when N = 1 and can use it to solve the rest.

Both recursion and induction require a similar conceptual approach involving a "leap of faith." In writing a recursive program, we assume that the solution procedure will correctly handle any new subproblems that arise, even if we cannot see all of the details in that solution. This corresponds to making the inductive hypothesis—the point at which we assume that the formula is true for some unspecified number N. If the formula we are trying to prove is not correct, this assumption will be contrary to fact in the general case. Nonetheless, the proof technique requires us to make that assumption and hold to it until we either complete the proof or establish a contradiction.

To illustrate this, suppose that we are attempting to prove the rather dubious proposition that "all positive integers are odd." As far as the base step is concerned, everything seems fine; the number 1 is indeed odd. To continue with the proof, we still begin by making the assumption that N is odd, for some unspecified value of N. The proof does not fall apart until we use that assumption in an attempt to prove that N+1 is also odd. By the laws of arithmetic, N+1 is even whenever N is odd, and we have discovered a contradiction to the original assumption.

In the case of both recursion and induction, we have no a priori reason to believe in the truth of our assumption. If it is valid, then the program or formula will operate correctly through all its levels. However, if there is an error in the recursive decomposition or a flaw in the inductive proof, the entire structure breaks down—the domino chain is broken. This faith in the correctness of something in which we as yet have no confidence is reminiscent of Coleridge when he describes "that willing suspension of disbelief for the moment, that constitutes poetic faith." Such faith is as important to the mathematician and the programmer as it is to the poet.

2-2 Computational Complexity

A few years ago, before the advent of the video arcade and the home computer, computer games were a relatively rare treat to be found at your local science or children's museum. One of the most widely circulated was a simple Guess-the-Number game which was played as follows:

> **Hi! Welcome to Guess-the-Number! I'm thinking of
> a number in the range 1 to 100. Try to guess it.
> What is your guess? 20
> That's too small, try again.**

> **What is your guess? 83**
> **That's too big, try again.**

The game continued in this fashion, accepting new guesses from the player, until the computer's secret number was discovered.

> **What is your guess? 37**
> **Congratulations! You got it in 12 guesses!**

For children who were happy to spend a little extra time with one of these games, twelve guesses hardly seemed excessive. Eventually, however, even relatively young players would discover that they could do a little better than this by exploiting a more systematic strategy.

To develop such a strategy, the central idea is that each guess must narrow the range to be searched as quickly as possible. This is accomplished by choosing the value closest to the middle of the available range. For example, the original problem can be expressed in English as

> Guess a number in the range 1 to 100.

If we guess 50 and discover that it is too large, we reduce this to the problem

> Guess a number in the range 1 to 49.

This has the effect of reducing the original problem to an identical subproblem in which the number is limited to a more restricted range. Eventually, we must guess the correct number, since the range will get smaller and smaller until only a single possibility remains.

In the language of computer science, this algorithm is called *binary search* and is an example of the recursive "divide-and-conquer" strategy presented in Chapter 1.* For this problem, binary search seems to work reasonably well. On the other hand, it is certainly not the only possible approach. For example, when asked to find a number in the range 1 to 100, we could certainly just ask a series of questions of the form

> Is it 1?
> Is it 2?
> Is it 3?

and so forth. We are bound to hit it eventually, after no more than 100 guesses. This algorithm is called *linear search* and is used quite frequently in computer science to find a value in an unordered list.

---

*As an algorithmic technique, binary search is of enormous practical importance to computer science and can be applied to many problems that are more exciting than the Guess-the-Number game. Nonetheless, Guess-the-Number provides an excellent setting for examining the algorithmic properties of the binary search technique.

Intuitively, we have a sense that the binary search mechanism is a better approach to the Guess-the-Number game, but we are not sure how much better it might be. In order to have some standard for comparison, we must find a way to measure the efficiency of each algorithm. In computer science, this is most often accomplished by calculating the *computational complexity* of the algorithm, which expresses the number of operations required to solve a problem as a function of the size of that problem.

The idea that computational complexity includes a consideration of problem size should not come as much of a surprise. In general, we expect that larger problems will require more time to solve than smaller ones. For example, guessing a number in the 1 to 1000 range will presumably take more time than the equivalent problem for 1 to 100. But how much more? By expressing efficiency as a relationship between size (usually represented by N) and the number of operations required, complexity analysis provides an important insight into how a change in N affects the required computational time.

As of yet, however, the definition of complexity is somewhat less precise than we might like, since the definition of an algorithmic operation depends significantly on the presentation of the algorithm involved. For a program that has been prepared for a particular computer system, we might consider each machine-language instruction as a primitive operation. In this context, counting the number of operations corresponds precisely to counting the number of instructions executed. Unfortunately, this would result in a measure of complexity which varies from machine to machine.

Alternatively, we can adopt the more informal definition of an "operation" as a simple conceptual step. In the case of the number-guessing problem, the only operation we perform is that of guessing a number and discovering how that number stands in relation to the value we are trying to discover. Using this approach, our measure of complexity is therefore the number of guesses required.

For many algorithms, the number of operations performed is highly dependent on the data involved and may vary widely from case to case. For example, in the Guess-the-Number game, it is always possible to "get lucky" and select the correct number on the very first guess. On the other hand, this is hardly useful in estimating the overall behavior of the algorithm. Usually, we are more concerned with estimating the behavior in (1) the *average* case, which provides some insight into the typical behavior of the algorithm, and (2) the *worst* case possible, which provides an upper bound on the required time.

In the case of the linear search algorithm, each of these measures is relatively easy to analyze. In the worst possible case, guessing a number in the range 1 to N might require a full N guesses. In the specific example involving the 1 to 100 range, this occurs if the number were exactly 100. To compute the average case, we must add up the number of guesses required for each possibility and divide that total by N. The number 1 is found on the first guess, 2 requires two guesses, and so forth, up to N, which requires N guesses. The sum of these possibilities is then

$$1 + 2 + 3 + \cdots + N = \frac{N(N+1)}{2}$$

Dividing this by N gives the average number of guesses, which is

$$\frac{N+1}{2}$$

The binary search case takes a little more thought but is still reasonably straightforward. In general, each guess we make allows us to reduce the size of the problem by a factor of two. For example, if we guess 50 in the 1 to 100 example and discover that our guess is low, we can immediately eliminate the values in the 1 to 50 range from consideration. Thus, the first guess reduces the number of possibilities to N/2, the second to N/4, and so on. Although, in some instances, we might get lucky and guess the exact value at some point in the process, the worst case will require continuing this process until there is only one possibility left. The number of steps required to accomplish this is illustrated by the diagram

$$N / 2 / 2 / 2 \cdots / 2 = 1$$

$$\underbrace{\qquad\qquad\qquad}_{k \text{ times}}$$

where k indicates the number of guesses required. Simplifying this gives

$$\frac{N}{2^k} = 1$$

or

$$N = 2^k$$

Since we want an expression for k in terms of the value of N, we must use the definition of logarithms to turn this around.

$$k = \log_2 N$$

Thus, in the worst case, the number of steps required to guess a number through binary search is the base-2 logarithm of the number of values.* In the average case, we can expect to find the correct value using one less guess (see Exercise 2-5).

---

*In analyzing the complexity of an algorithm, we frequently will make use of logarithms which, in almost all instances, are calculated using 2 as the logarithmic base. For the rest of this text, we will follow the standard convention in computer science and use "log N" to indicate base 2 logarithms without explicitly writing down the base.

Estimates of computational complexity are most often used to provide insight into the behavior of an algorithm as the size of the problem grows large. Here, for example, we can use the worst-case formula to create a table showing the number of guesses required for the linear and binary search algorithms, respectively:

N	Linear search	Binary search
10	10	4
100	100	7
1,000	1,000	10
10,000	10,000	14
100,000	100,000	17
1,000,000	1,000,000	20

This table demonstrates conclusively the value of binary search. The difference between the algorithms becomes increasingly pronounced as N takes on larger values. For ten values, binary search will yield the result in no more than four guesses. Since linear search requires ten, the binary search method represents a factor of 2.5 increase in efficiency. For 1,000,000 values, on the other hand, this factor has increased to one of 50,000. Surely this represents an enormous improvement.

In the case of the search algorithms presented above, the mechanics of each operation are sufficiently simple that we can carry out the necessary computations in a reasonably exact form. Often, particularly when analyzing specific computer programs, we are forced to work with approximations rather than exact values. Fortunately, those approximations turn out to be equally useful in terms of predicting relative performance.

For example, consider the following nested loop structure in Pascal:

```
FOR I := 1 TO N DO
   FOR J := 1 TO I DO
      A[I,J] := 0
```

The effect of this statement, given a two-dimensional array A, is to set each element on or below the main diagonal to zero. At this point, however, our concern is not with the purpose of the program but with its computational efficiency. In particular, how long would this program take to run, given a specific matrix of size N?

As a first approximation to the running time, we should count the number of times the statement

$$A[I,J] := 0$$

is executed. On the first cycle of the outer loop, when I is equal to 1, the inner loop will be executed only once for $J = 1$. On the second cycle, J will run through both the values 1 and 2, contributing two more assignments. On the last cycle, J will range through all N values in the 1 to N range. Thus, the total number of assignments to some element $A[I,J]$ is given by the formula

$$1 + 2 + 3 + \cdots + N$$

Fortunately, we have seen this before in the discussion of mathematical induction and know that this may be simplified to

$$\frac{N(N + 1)}{2}$$

or, expressing the result in polynomial form,

$$\tfrac{1}{2}N^2 + \tfrac{1}{2}N$$

Although this count is accurate in terms of the number of assignments, it can be used only as a rough approximation to the total execution time since it ignores the other operations that are necessary, for example, to control the loop processes. Nonetheless, it can be quite useful as a tool for predicting the efficiency of the loop operation as N grows large. Once again, it helps to make a table showing the number of assignments performed for various values of N.

N	Assignments to A[I,J]
10	55
100	5,050
1,000	500,500
10,000	50,005,000

From this table, we recognize that the number of assignments grows much more quickly than N. Whenever we multiply the size of the problem by ten, the number of assignments jumps by a factor of nearly 100.

The table also illustrates another important property of the formula

$$\tfrac{1}{2}N^2 + \tfrac{1}{2}N$$

As N increases, the contribution of the second term in the formula decreases in importance. Since this formula serves only as an approximation, it was probably rather silly to write 50,005,000 as the last entry in the table. Certainly, 50,000,000 is close enough for all practical purposes. As long as N is relatively large, the first term will always be much larger than the second. Mathematically, this is often indicated by writing

$$N^2 >> N$$

The symbol ">>" is read as "dominates" and indicates that the term on the right is insignificant compared with the term on the left, whenever the value of N is sufficiently large. In more formal terms, this relationship implies that

$$\lim_{N \to \infty} \frac{N}{N^2} = 0$$

Since our principal interest is usually in the behavior of the algorithm for large values of N, we can simplify the formula and say that the nested loop structure runs in a number of steps roughly equal to

$$\tfrac{1}{2}N^2$$

Conventionally, however, computer scientists will apply yet another simplification here. Although it is occasionally useful to have the additional precision provided by the above formula, we gain a great deal of insight into the behavior of this algorithm by knowing that it requires a number of steps *proportional* to

$$N^2$$

For example, this tells us that if we double N, we should expect a fourfold increase in running time. Similarly, a factor of ten increase in N should increase the overall time by a factor of 100. This is indeed the behavior we observe in the table, and it depends only on the

$$N^2$$

component and not any other aspect of the complexity formula.

In computer science, this "proportional" form of complexity measure is used so often that it has acquired its own notational form. All the simplifications that were introduced above can be summarized by writing that the Pascal statement

```
FOR I := 1 TO N DO
    FOR J := 1 TO I DO
        A[I,J] := 0
```

has a computational complexity of

144.976 $O(N^2)$

This notation is read either as "big-O of N squared" or, somewhat more simply, "order N squared."

Formally, saying that a particular algorithm runs in time

$$O(f(N))$$

for some function f(N) means that, as long as N is sufficiently large, the time required to perform that algorithm is never larger than

$$C \times f(N)$$

for some unspecified constant C.

In practice, certain orders of complexity tend to arise quite frequently. Several common algorithmic complexity measures are given in Table 2-1. In the table, the last column indicates the name which is conventionally used to refer to algorithms in that class. For example, the linear search algorithm runs, not surprisingly, in linear time. Similarly, the nested Pascal loop is an example of a quadratic algorithm.

Table 2-1. Common Complexity Characteristics

Given an algorithm of this complexity	When N doubles, the running time	Conventional name
$O(1)$	Does not change	Constant
$O(\log N)$	Increases by a small constant	Logarithmic
$O(N)$	Doubles	Linear
$O(N \log N)$	Slightly more than doubles	N log N
$O(N^2)$	Increases by a factor of 4	Quadratic
$O(N^k)$	Increases by a factor of 2^k	Polynomial
$O(\alpha^N)$ $\alpha > 1$	Depends on α, but grows very fast	Exponential

The most important characteristic of complexity calculation is given by the center column in the table. This column indicates how the performance of the algorithm is affected by a doubling of the problem size. For example, if we double the value of N for a quadratic algorithm, the run time would increase by a factor of four. If we were somehow able to redesign that algorithm to run in time N log N, doubling the size of the input data would have a less drastic effect. The new algorithm would still require more time for the larger problem, but the increase would be only a factor of two over the smaller problem (plus some constant amount of time contributed by the logarithmic term). If N grows even larger, this constitutes an enormous savings in time.

In attempting to improve the performance of almost any program, the greatest potential savings come from improving the algorithm so that its complexity bound is reduced. "Tweaking" the code so that a few instructions are eliminated along the way can only provide a small percentage increase in performance. On the other hand, changing the algorithm offers an unlimited reduction in the running time. For small problems, the time saved may be relatively minor, but the percentage savings grows much larger as the size of the problem increases. In many practical settings, algorithmic improvement can reduce the time requirements of a program by factors of hundreds or thousands—clearly an impressive efficiency gain.

Bibliographic Notes

One of the best discussions of mathematical induction and how to use it is contained in Solow's *How to Read and Do Proofs* [1982]. The classic text in this area is Polya [1957]. A more formal discussion of asymptotic complexity and the use of the big-O notation may be found in Knuth [1973] or Sedgewick [1983].

The inductive argument used in Exercise 2-6 to "prove" that all horses are the same color is adapted from Joel Cohen's essay "On the Nature of Mathematical Proofs" [1961].

Exercises

2-1. Prove, using mathematical induction, that the sum of the first N even integers is given by the formula

$$N^2 + N$$

How could you predict this expression using the other formulae developed in this chapter?

2-2. Establish that the following formulae are correct using mathematical induction:

(a) $1 + 2 + 4 + 8 + \cdots + 2^{N-1} = 2^N - 1$

(b) $1 + 3 + 9 + 27 + \cdots + 3^N = \dfrac{3^{N+1} - 1}{2}$

(c) $1 \times 1 + 2 \times 2 + 3 \times 4 + 4 \times 8 + \cdots + N \times 2^{N-1} = (N-1) 2^N + 1$

2-3. The year: 1777. The setting: General Washington's camp somewhere in the colonies. The British have been shelling the Revolutionary forces with a large cannon within their camp. You have been assigned a dangerous reconnaissance mission—to infiltrate the enemy camp and determine the amount of ammunition available for that cannon.

Fortunately for you, the British (being relatively neat and orderly) have stacked the cannonballs into a single pyramid-shaped stack. At the top is a single cannonball resting on a square of four cannonballs, which is itself resting on a square of nine cannonballs, and so forth. Given the danger in your situation, you only have a chance to count the number of layers before you escape back to your own encampment.

Using mathematical induction, prove that, if N is the number of layers, the total number of cannonballs is given by

$$\frac{N(N+1)(2N+1)}{6}$$

2-4. Assuming that N is an integer which defines the problem size, what is the order of each of the program fragments shown below:

(a) **K := 0;**
   FOR I := 1 TO 1000 DO
     K := K + I * N;

(b) **K := 0;**
   FOR I := 1 TO N DO
     FOR J := I TO N DO
       K := K + I * J;

(c) **K := 0;**
   WHILE N > 0 DO
     BEGIN
       N := N DIV 2;
       K := K + 1
     END;

Assuming that the statements within the loop body take the same amount of time, for what values of N will program (a) run more quickly than program (c)?

2-5. [For the mathematically inclined] In trying to locate a number by binary search, it is always possible that it will require many fewer guesses than would be required in the worst possible case. For example, if 50 were in fact the correct number in the 1 to 100 range, binary search would find it on the very first guess. To determine the average-case behavior of binary search, we need to determine the expected value of the number of guesses over the entire possible range.

Assuming that there are N numbers in the complete range, we know that only one of them (specifically, the number at the center of the range) will be guessed on the very first try. Two numbers will be guessed in two tries, four numbers in three tries, and so forth. Thus, the average number of guesses required is given by the formula

$$\frac{1 \times 1 + 2 \times 2 + 3 \times 4 + 4 \times 8 + \cdots + G \times 2^{G-1}}{N}$$

where G is the maximum number of guesses that might be required, which in this case is simply log N.

Using the expression from Exercise 2-2(c), simplify this formula. As N grows large, what does this approach?

2-6. Mathematical induction can have its pitfalls, particularly if one is careless. For example, the following argument appears to be a proof that all horses are the same color. What is wrong here? Is there really no horse of a different color?

Preliminary   Define a set of horses to be *monochromatic* if all the horses in the set have the same coloration.

Conjecture   Any set of horses is monochromatic.

Technique   Proof by induction on the number of horses.

Base step   Any set of one horse is monochromatic, by definition.

Induction   Assume that any set of N horses is monochromatic. Consider a set of N + 1 horses. That can be divided into smaller subsets in several ways. For example, consider the division indicated in the following diagram:

$$\underbrace{H_1 H_2 \cdots H_N}_{A} \underbrace{H_{N+1}}_{A'}$$

The subset labeled A in this diagram is a set of N horses and is therefore monochromatic by the inductive hypothesis.

Similarly, if we divide the complete set as follows:

$$\underbrace{H_1 \underbrace{H_2 \cdots H_N H_{N+1}}_{B}}_{}$$
$$\quad B'$$

we get a subset B which is also monochromatic.

Thus, all the horses in subset A are the same color as are all the horses in subset B. But H_2 is in both subsets A and B, which implies that both subsets must contain horses of the *same* color.

Recursive Functions

To iterate is human, to recurse divine.—Anonymous

In Chapter 1, recursion was defined as a solution technique that operates by reducing a large problem to simpler problems of the same type. As written, this is an abstract definition and describes the recursive process in terms which do not necessarily imply the use of a computer. In order to understand how recursion is applied in the context of a programming language, this definition must be recast in a more specific form.

In formulating this new definition, we must first find a way to represent an algorithm so that it solves not only a specific problem but also any sub-problems that are generated along the way. Returning to the fund-raising example from Chapter 1, it is not sufficient to write a single procedure which collects 1000 dollars. Instead, the recursive implementation must correspond to the more general operation of raising N dollars, where N is a *parameter* that changes during the course of the recursive solution.

Given that we need parameters to define a specific instance of the problem, recursive solutions are usually implemented as *subroutines** whose arguments convey the necessary information. Whenever a recursive routine breaks a large problem down into simpler subproblems, it solves those subproblems by *calling the original routine* with new arguments, updated to reflect each subproblem.

During the course of a recursive solution, old problems are set aside as new subproblems are solved; when these are completed, work must continue on the problems that were previously deferred. As discussed in Chapter 1, this requires the system to maintain a list of unfinished tasks.

In order for recursion to be a useful tool, the programming language must ensure that this bookkeeping is performed automatically so that the programmer need not be bothered with the details. Unfortunately, many of the older and more established languages, such as FORTRAN, were designed when the im-

*In this context, "subroutine" is used as a generic term to refer to either of the two program types supported by Pascal: procedures and functions.

portance of recursive programming techniques was not so widely recognized and do not "support recursion" in this way. In these languages, recursive techniques are much more difficult to use and must be simulated using the techniques described in Chapter 11.

3-1   Functional vs. Procedural Recursion

Like most modern programming languages, Pascal provides two distinct sub-routine types: *procedures* and *functions*. In many respects, the two are nearly identical from a technical point of view. They share a common syntactic structure and are usually represented by the same sequence of instructions in the underlying machine-language implementation. The principal difference is conceptual. Procedures are used at the statement level and are executed for their *effect*. Functions are used as part of an expression and are executed for the *value* they return.

In languages which support recursion, both functions and procedures can be written to take advantage of recursive techniques. Recursive functions operate by defining a particular function in terms of the values of that same function with simpler argument values. Recursive procedures tend to be oriented toward problem solving and therefore correspond more closely to the abstract examples of recursion given in Chapter 1.

Even though recursive procedures tend to provide a better conceptual model for understanding how a problem can be subdivided into simpler components, recursive functions are traditionally introduced first. There are two principal advantages to this mode of presentation. First, there are several mathematical functions (such as the factorial and Fibonacci functions described below) which are particularly elegant when expressed in a recursive form. Second, the recursive character of these examples follows directly from their mathematical definition and is therefore more clear.

3-2   Factorials

In word games such as the SCRABBLE Crossword Game,* the play consists of rearranging a set of letters to form words. For example, given the seven letters

T   I   R   N   E   G   S

we can form such words as RIG, SIRE, GRINS, INSERT, or RESTING. Often, there is a considerable advantage in playing as long a word as possible. In

*SCRABBLE is a registered trademark of the Selchow and Righter Company of Bay Shore, New York.

SCRABBLE, for example, playing all seven tiles in the same turn is rewarded with a bonus of 50 points, making RESTING (or STINGER) a particularly attractive play.

In Chapter 6, we will turn our attention to the more interesting problem of generating all the possible arrangements, but we already have the necessary tools to consider a somewhat simpler question. Given a set of seven distinct letters, how many different arrangements must we check to discover all possible seven-letter words?

On the whole, this is not too difficult a task. In constructing all possible arrangements, there are seven different ways to select the starting letter. Once that has been chosen, there are six ways to choose the next letter in sequence, five to choose the third, and so on, until only one letter is left to occupy the last position. Since each of these choices is independent, the total number of orderings is the product of the number of choices at each position. Thus, given seven letters, there are

$$7 \times 6 \times 5 \times 4 \times 3 \times 2 \times 1$$

arrangements, which works out to be 5040.

In mathematical terms, an arrangement of objects in a linear order is called a *permutation*. Given a set of N distinct objects, we can calculate the number of permutations of that set by applying much the same analysis. There are N ways of choosing the first object, N–1 for the second, N–2 for the third, and so forth down to 1. Thus, the total number of permutations is given by the formula

$$N \times (N–1) \times (N–2) \times \ldots \times 1$$

This number is defined to be the *factorial* of N and is usually written as N! in mathematics. Here, to ease the transition to Pascal, factorials will be represented in a functional form, so that the appropriate definition is

$$FACT(N) = N \times (N–1) \times (N–2) \times \ldots \times 1$$

Our task here is to write a Pascal function which takes an integer N as its argument and returns N! as its result. Given this definition of the problem, we expect the function header line to be

FUNCTION FACT(N : INTEGER) : INTEGER;

All that remains is writing the necessary code that implements the factorial computation.

To gain some intuition about the behavior of the function **FACT**, it helps to construct a table for the first few values of N, including the case N = 0, for which the factorial is defined to be 1 by mathematical convention.

$$FACT(0) = \qquad\qquad\qquad 1$$
$$FACT(1) = \qquad\qquad 1 = 1$$
$$FACT(2) = \qquad\quad 2 \times 1 = 2$$
$$FACT(3) = \quad 3 \times 2 \times 1 = 6$$
$$FACT(4) = 4 \times 3 \times 2 \times 1 = 24$$

At this point, there are two distinct approaches to this problem. The *iterative* approach consists of viewing the factorial computation as a series of multiplications. To calculate **FACT(4)**, for example, start with 1 and multiply it by each of the numbers up to 4. This gives rise to a program such as

```
FUNCTION FACT(N : INTEGER) : INTEGER;
   VAR
   I, PRODUCT : INTEGER;
   BEGIN
     PRODUCT := 1;
     FOR I := 1 TO N DO
       PRODUCT := I * PRODUCT;
     FACT := PRODUCT
   END;
```

Proceeding recursively leads to a rather different implementation of the FACTORIAL function. In examining the factorial table, it is important to observe that each line contains exactly the same product as the previous one, with one additional factor. This makes it possible to represent the table in a simpler form:

$$FACT(0) = 1$$
$$FACT(1) = 1 \times FACT(0)$$
$$FACT(2) = 2 \times FACT(1)$$
$$FACT(3) = 3 \times FACT(2)$$
$$FACT(4) = 4 \times FACT(3)$$

or, in more general terms,

$$FACT(N) = N * FACT(N-1)$$

This is a recursive formulation because it expresses the calculation of one factorial in terms of the factorial of a smaller integer. To calculate **FACT(50)**, we simply take 50 and multiply that by the result of calculating **FACT(49)**. Calculating **FACT(49)** is accomplished by multiplying 49 by the result of **FACT(48)**, and so forth.

As in any recursive problem, a simple case is necessary to ensure that the calculation will terminate at some point. Since 0 is conventionally used as the base of the factorial table in mathematics, this suggests the following definition:

$$FACT(N) = \begin{cases} 1, & \text{if } N = 0 \\ N \times FACT(N-1), & \text{otherwise} \end{cases}$$

In a language like Pascal that supports recursion, it is easy to take the recursive definition of factorial and transform it into the necessary Pascal code:

```
FUNCTION FACT(N : INTEGER) : INTEGER;
   BEGIN
      IF N = 0 THEN
         FACT := 1
      ELSE
         FACT := N * FACT(N-1)
   END;
```

Note that the program and the abstract definition have almost exactly the same form. This represents a significant advantage if we are concerned with program clarity. In particular, if we are convinced of the correctness of a recursive mathematical definition, the similarity in form between that definition and the corresponding program makes the correctness of the program that much easier to establish.

As in the case of any recursive implementation, we can look at the process of computing factorials from two philosophical perspectives. The holistic observer examines the mathematical definition along with the corresponding program and then walks away quite satisfied. For the reductionist, it is necessary to follow the program at a much more detailed level.

For example, in order to calculate **FACT(6)**, the function evaluates its argument, discovers that it is not zero, and proceeds to the **ELSE** clause. Here it discovers that it needs to compute the expression

$$N * FACT(N-1)$$

and return this as the value of the function. Given that $N = 6$ here, this means that the solution at this level is simply

$$6 * FACT(5)$$

Evaluating the second term in this expression, however, requires making a recursive call to the **FACT** function. At this point, the program "remembers" (by making the appropriate notations on the list of unfinished tasks) that it is in the process of multiplying 6 by the result of a new factorial computation, specifically **FACT(5)**. Repeating this process leads to a cascading set of factorial calculations which can be represented diagrammatically as follows:

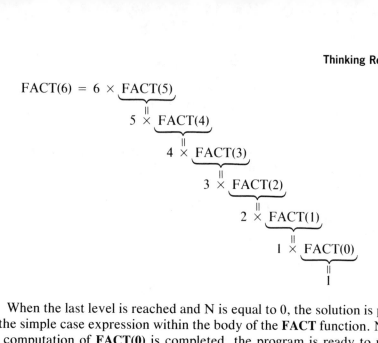

When the last level is reached and N is equal to 0, the solution is provided by the simple case expression within the body of the **FACT** function. Now that the computation of **FACT(0)** is completed, the program is ready to return to the previous level of call. By referring back to the list of unfinished tasks, we discover that we were in the middle of computing

1 * FACT(0)

Given the new information that **FACT(0)** is 1, we can complete this operation and return to the next higher level. From here, we simply climb back up through the sequence of multiplications, eventually computing 720 as the final answer.

3-3   The Fibonacci Sequence

In his text on computation (*Liber Abbaci*) written in 1202, the Italian mathematician Leonardo Fibonacci includes an exercise which may be considered as one the earliest excursions into the field of population biology. Rendered in more modern terms, the Fibonacci problem has the following form:

> Assume that the reproductive characteristics of rabbits can be defined by the following rules:
>
> **1.** Each pair of fertile rabbits produces a new pair of offspring each month.
> **2.** Rabbits become fertile in their second month of life.
> **3.** Old rabbits never die.
>
> If a single pair of newborn rabbits is introduced at the end of January, how many pairs of rabbits will there be at the end of a year?

We can begin this calculation by tracing through the history of the rabbit colony during the first few months. At the end of the first month, we have only our original pair of rabbits. Since these are newborn rabbits, February is an unproductive month, and we still have only one pair of rabbits at the end of month 2. In March, however, our original pair is now fertile, and a new pair of rabbits is born, increasing the colony's population (counting by pairs) to two. In April, the original pair goes right on producing, but the March rabbits (distant relatives of the March Hare) are as yet too young. This gives us three pairs of rabbits at the end of month 4.

From here (as rabbits are wont to do), things begin to grow rather explosively. In May, both the original pair and the March rabbits are in the rabbit-making business, and two new pairs of rabbits are born. Given that there were three before, this results in a total of five. We can summarize the information to date in a simple table:

At the end of month	Number of rabbit pairs
1	1
2	1
3	2
4	3
5	5

From here, we can calculate the remaining entries by making an important observation. The rabbits that are part of the population at the end of any month come from two distinct sources. First of all, since old rabbits never die, all the rabbits that were around in the previous month are still around. In addition, every pair of rabbits that is old enough to reproduce gives birth to a new pair. This number is equivalent to the number of pairs that were alive *two* months ago, since all of these must, by this time, be capable of reproduction. This observation gives rise to the following computational rule:

The number of rabbit pairs at the end of month N is equal to the sum of

1. The number of pairs at the end of month N-1
2. The number of pairs at the end of month N-2

If we use the notation **FIB(N)** to denote the number of rabbit pairs at the end of month N (where FIB is chosen in honor of Fibonacci), we can rephrase this discovery in a more compact form:

$$\textbf{FIB(N)} \;=\; \textbf{FIB(N–1)} \;+\; \textbf{FIB(N–2)}$$

An expression of this type, which defines a particular element of a sequence in terms of earlier elements, is called a *recurrence relation* and is part of a recursive definition.

The recurrence relation makes it possible to complete the table and answer Fibonacci's question concerning the number of rabbits at the end of the year. To compute each entry in the right-hand column, we simply add the two previous entries.

N	FIB(N)
1	1
2	1
3	2
4	3
5	5
6	8
7	13
8	21
9	34
10	55
11	89
12	144

In mathematics, this set of numbers is called the *Fibonacci sequence* and turns up in a surprising number of practical contexts.

To complete the recursive definition of the Fibonacci sequence, however, we need to specify certain simple cases in order to ensure that the recursion terminates. In this example, defining

$$\textbf{FIB(1)} \ = \ \textbf{1}$$

is not sufficient, since it leaves us unable to compute **FIB(2)**. From the recurrence relation, **FIB(2)** is calculated as

$$\textbf{FIB(2)} \ = \ \textbf{FIB(1)} \ + \ \textbf{FIB(0)}$$

To accomplish this addition, both values on the right-hand side must be defined. In order for **FIB(2)** to have the correct value, **FIB(0)** must be defined as

$$\textbf{FIB(0)} \ = \ \textbf{0}$$

This makes it possible to complete the formal definition

$$FIB(N) \ = \ \begin{cases} 0, & \text{if } N = 0 \\ 1, & \text{if } N = 1 \\ FIB(N-1) + FIB(N-2) & \text{otherwise} \end{cases}$$

As in the case of the factorial function, a definition of this sort can be turned into a recursive program in an entirely straightforward way. Taking advantage of the fact that **FIB(N)=N** for both of the simple cases **N=0** and **N=1**, we can implement this function in Pascal as

```
FUNCTION FIB(N : INTEGER) : INTEGER;
   BEGIN
      IF (N = 0) OR (N = 1) THEN
         FIB := N
      ELSE
         FIB := FIB(N-1) + FIB(N-2)
   END;
```

On the other hand, this solution technique is hardly ideal. To its credit, we should note that this implementation is both correct and concise; moreover, its correctness is easily justified by appealing to the corresponding mathematical definition. The problem here is one of efficiency.

To understand why this implementation is considerably less efficient than we would like, it helps to examine the complete computation sequence for a simple value, such as **N** = 4. The complete computation involves the following recursive decomposition:

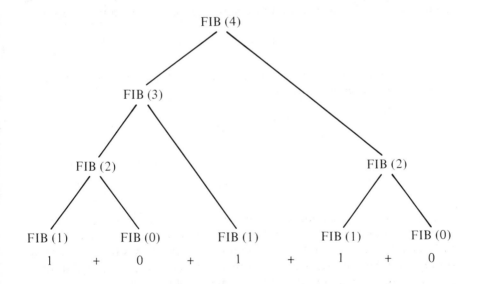

This diagram begins to illustrate the source of the efficiency problem. In calculating **FIB(4)** by this approach, **FIB(2)** and **FIB(0)** are each calculated twice, and **FIB(1)** is calculated three times. As **N** increases, the re-dundancy of computation grows even worse. For example, to calculate **FIB(5)**, we would not only be forced to duplicate the entire calculation above, but we would also recalculate from scratch another copy of **FIB(3)**.

By counting the number of additions required to calculate **FIB(N)** for small values of **N,** we can begin to derive some sense of the computational complexity of this algorithm.

N	FIB(N)	Additions required
0	0	0
1	1	0
2	1	1
3	2	2
4	3	4
5	5	7
6	8	12
7	13	20
8	21	33

Clearly, as **N** grows, the number of additions required to compute **FIB(N)** by this mechanism increases fairly quickly. Although we as yet have only circumstantial evidence for this conjecture (see Exercise 3-7), it appears that the number of additions required to compute **FIB(N)** is given by the formula

$$\textbf{FIB(N + 1)} - \textbf{1}$$

By applying some additional mathematics, we can use this result to establish that the complexity order for this algorithm is

$$O(\varphi^N)$$

where

$$\varphi \approx 1.618034$$

The details of this derivation are largely irrelevant to understanding the computational complexity that this implies and have been left to the mathematically inclined reader as Exercise 3-9. The important observation is that this approach to computing the Fibonacci sequence results in an *exponential* algorithm and is, as such, impractical for all but the smallest values of N.

The problem is not inherent in the calculation of the Fibonacci sequence itself, but is instead a property of the solution technique. A much more efficient method of computing the Fibonacci sequence is illustrated by the following iterative version of **FIB:**

```
FUNCTION FIB(N : INTEGER) : INTEGER;
VAR
  A, B, C, I : INTEGER;
BEGIN
  IF (N = 0) OR (N = 1) THEN
    FIB := N
  ELSE
    BEGIN
      A := 0;
      B := 1;
      FOR I := 2 TO N DO
        BEGIN
          C := A + B;
          A := B;
          B := C
        END;
      FIB := C
    END
END;
```

On the negative side, this program is considerably longer and requires several temporary variables. Moreover, the correspondence between this algorithm and the mathematical formulation is not immediately clear. On the other hand, by examining the structure of the **FOR** loop in the program, we can determine that it does run in time proportional to the value of N. We have thus transformed an exponential implementation into a linear one. This represents a dramatic improvement in efficiency.

After looking at the Fibonacci example, there is a certain temptation to condemn recursive strategies as inherently inefficient. From the circumstantial evidence above, this seems entirely reasonable, but it is actually rather unfair. The culprit here is not recursion, but the way recursion is applied. The recurrence relation

$$FIB(N) = FIB(N-1) + FIB(N-2)$$

is intended to define the mathematical properties of this sequence and need not be regarded as an appropriate computational technique. This question is examined further in the exercises.

Bibliographic Notes

Almost all programming texts which include a discussion of recursion approach the topic from the standpoint of recursive functions. Thus, reasonable sources of additional material include such texts as Cooper and Clancy [1982], Grogono [1984], and Tenenbaum and Augenstein [1981]. Further information about the mathematical properties of the Fibonacci sequence is available in Knuth [1973].

Exercises

3-1. Ostensibly for reasons of efficiency, Pascal contains no operator which will raise a number to a given power. Assuming that the exponent K is always a nonnegative integer, write a recursive function **POWER(X, K)** which raises the real value X to the K power.

3-2. Without using the explicit formula, write a recursive function which computes the answer to the "cannonball" problem from Exercise 2-3.

3-3. As illustrated in Chapter 2, the sum of the first N odd integers has the value N squared. In particular, this makes it possible to write a recurrence relation that defines the square of an integer in terms of the square of the previous integer.

Write out this recurrence relation, and, by adding an appropriate simple case, complete the definition of the Pascal function

FUNCTION SQUARE(N : INTEGER) : INTEGER;

which squares its argument.

3-4. The greatest common divisor of two nonnegative integers X and Y is defined to be the the largest integer which divides evenly into both. For example, given the numbers 28 and 63, the greatest common divisor is 7. This problem was considered to be of great importance to classical mathematicians, including Euclid, whose *Elements* (book 7, proposition II) contains an elegant solution.

The essence of Euclid's discovery is that the greatest common divisor of X and Y must always be the same as that of Y and R, where R is the remainder of X divided by Y. By providing an appropriate simple case, use this relationship to develop a recursive coding of a function GCD which takes two integers and returns their greatest common divisor.

3-5. As noted in the text, the number of arrangements of the seven letters in a SCRABBLE rack can be determined by computing seven factorial. In playing SCRABBLE, however, one is rarely lucky enough to play all seven tiles at once and must often be content with finding a shorter word. This suggests a related question: Given a rack of seven distinct letters, how many different arrangements are possible using some smaller number of letters?

If we generalize this, we derive the problem of selecting an ordered set of K elements from a collection of N distinct objects. In mathematics, this is represented as the function

$$P(N,K)$$

which is interpreted as "the number of permutations of N objects taken K at a time."

By following through the analysis in the text, determine the recursive structure for **P(N,K)** and construct a Pascal function which implements this algorithm.

3-6. In working with permutations, the order in which the elements are selected is significant. Thus, "ABC" is not the same permutation as "CBA". In some applications, this order is irrelevant. For example, if we are merely interested in determining how many different sets of four letters can be chosen from the set {A, B, C, D, E, F, G}, we have a different type of problem.

Mathematically, selections which are made without regard to order are called *combinations*. Although other notations are also used, it is convenient to represent the number of combinations that can be selected from a set of **N** objects taken **K** at a time as the function

$$C(N,K)$$

There are many ways to calculate the value of **C(N,K)**. One of the most aesthetically interesting is to use the Pascal triangle (named for the mathematician, not the programming language):

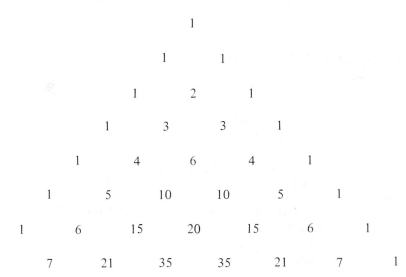

In this triangle, the left and right edges are always 1, and every remaining element is the sum of the two elements above it to the right and left.

If we use **N** to indicate a row in this table, and **K** a diagonal "column," then the entry at position **(N,K)** in the Pascal triangle corresponds to **C(N,K)**. This is made more explicit by labeling the diagram as shown:

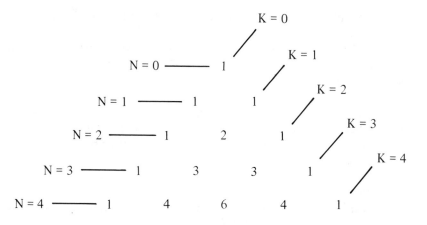

For example, this shows that **C(4,2)** has the value 6. This means that, given a set of four objects, there are six different ways to choose an unordered pair.

By considering the rule given above for constructing the Pascal triangle, write a recursive implementation of the function **C(N,K)**.

3-7. Using mathematical induction, prove that the number of additions required to calculate **FIB(N)** using the recursive algorithm given in the text is given by

$$FIB(N+1) \; - \; 1$$

3-8. Consider the following pair of function definitions:

```
FUNCTION G(N, A, B : INTEGER) : INTEGER;
   BEGIN
      IF N = 0 THEN
         G := A
      ELSE
         G := G(N-1, B, A + B)
   END;

FUNCTION F(N : INTEGER) : INTEGER;
   BEGIN
      F := G(N, 0, 1)
   END;
```

Compute the values of **F(0)**, **F(1)**, **F(2)**, and so forth up to **F(5)**. What mathematical function does **F** compute? In terms of **N**, how many additions are performed during this calculation? Given this, what is the complexity order of this algorithm?

Note that all the work here is performed by the recursive function **G** and that **F** exists only to set the correct initial values of the parameters **A** and **B**. This is a common technique in recursive programming and will reappear in Chapter 7.

3-9. [Mathematically oriented] The constant φ referred to in the text is called the "golden ratio" and has considerable significance, not only in mathematics but also in art. During the Renaissance, φ was often called the "divine proportion" and was considered to have important aesthetic value when used as the ratio between different elements within an artistic representation. Mathematically, φ is defined as the number which satisfies the proportionality equation:

$$\frac{1}{\varphi} = \frac{\varphi}{\varphi + 1}$$

Using cross-multiplication, we can rearrange the terms so that φ is expressed in quadratic form:

$$\varphi^2 - \varphi - 1 = 0$$

From here, we can use the quadratic formula to find the two solutions to this equation:

$$\varphi = \frac{1 + \sqrt{5}}{2}$$

and

$$\hat{\varphi} = \frac{1 - \sqrt{5}}{2}$$

The relationship between the golden ratio and the Fibonacci numbers was demonstrated by the mathematician A. de Moivre, who discovered that the Nth Fibonacci number can be expressed using the formula

$$\frac{(\varphi^N - \hat{\varphi}^N)}{\sqrt{5}}$$

Using mathematical induction, prove that this formula is correct. To do this, it is necessary to prove that this formula gives the correct values of **FIB(0)** and **FIB(1)** and then show that

$$\textbf{FIB(N)} \;=\; \textbf{FIB(N--1)} \;+\; \textbf{FIB(N--2)}$$

for all larger values. In working through the details of the proof, it is often useful to remember that both φ and $\hat{\varphi}$ satisfy the equation

$$\varphi^2 = \varphi + 1$$

The Procedural
Approach

4

For each of the functions presented in Chapter 3, the necessary recursive decomposition into simpler instances follows directly from the mathematical definition. For example, the Fibonacci sequence is most clearly defined by the recurrence relation

$$\text{FIB}(N) = \text{FIB}(N-1) + \text{FIB}(N-2)$$

Given this definition, the recursive implementation seems the most natural one, even if it is not the most efficient.

Mathematical formalism, such as the use of recurrence relations, is more easily applied to the domain of recursive functions than to recursive procedures. Intuitively, functions are mathematical entities, while procedures are more algorithmic in character. Thus, procedures are less likely to have a simple mathematical formulation, making the necessary recursive decomposition of the problem somewhat harder to find. Learning to apply recursive techniques in a procedural context is the essence of thinking recursively and requires a relatively abstract view of procedures themselves.

To the reductionist, a procedure call is simply a transfer of control that remembers the address of the instruction making that call. When the operation of the procedure is complete, control returns to that saved address, allowing the calling program to continue from the point at which the call was made. From a more holistic perspective, however, the procedure acts principally as a mechanism for suppressing irrelevant detail.

For example, many programming applications consist of (1) reading a set of data, (2) analyzing some aspect of that data, and (3) generating reports that present the results of the analysis. To the extent that these components represent independent phases of the total solution, it is useful to separate them algorithmically by making each component a separate procedure. Adopting this approach leads to a main program of the following form:

```
                    BEGIN
                        READDATA;
                        ANALYZEDATA;
                        PRINTREPORT
                    END.
```

The advantage of this programming style is that it enables us, as readers of the program, to consider the algorithm at varying levels of detail. At the most general level, for example, we can understand how the principal components fit together by looking at the main program in isolation. Beyond that, the individual components of the task can be examined in any order, and only when additional detail is required.

When we think about programming languages from the perspective of human linguistics, writing a procedure has the effect of defining a new imperative verb. When we write the procedure **READDATA,** for example, the word

READDATA

is added to our command vocabulary and can be used in the same fashion as any of the built-in operations. Once the debugging is complete and we develop the necessary level of confidence in **READDATA**'s correctness, we begin to ignore its internal detail and think of it as a new high-level operation.

In a sense, the definition of a procedure establishes a boundary between two different levels of abstraction. When the procedure is defined, we are principally concerned with the details of its operation. When the procedure is called, however, we are on the opposite side of the fence and consider only the effect of that call.

For example, returning to the problem of generating "computer art" outlined in Chapter 1, we might define the procedure **DRAWCUBIST** (with appropriate arguments to specify the size and location of the canvas) to be the phrase

"Draw a Cubist painting on the indicated canvas"

When we call this from the main program, we ignore the details and accept this definition at its abstract level.

When we write the body of **DRAWCUBIST,** we need to consider its internal operation, but must limit our attention to a single level of detail. As in the case of any recursive program, we will eventually reach a point in **DRAWCUBIST** at which we have isolated a subproblem in precisely the same form as the original. Here, we will again need to perform the operation

"Draw a Cubist painting on the indicated canvas"

for a canvas with smaller dimensions. Fortunately, we have a procedure which accomplishes precisely this task, and we can simply use a call to **DRAWCUBIST** with the appropriate set of arguments. Since this is a procedure call, we consider only the effect of the procedure rather than its internal details.

Admittedly, adopting this conceptual perspective is a bit difficult. Intuitively, our minds rebel a bit at the idea of using (especially in such a trusting way) a procedure we have not yet finished writing. Nonetheless, doing just that is the essential tactical step in writing recursive programs and constitutes the "recursive leap of faith." Trying to analyze further than this only clouds the picture with irrelevant detail. Metaphorically, looking inside to ensure that the recursive call does the job is very much like Orpheus looking back to make sure that Eurydice is still following—sometimes you just have to believe.

4-1   Numeric Output

To anyone working in the environment of a modern high-level language, the idea of calling a subroutine without first understanding the details of its internal operation should hardly be unfamiliar. In the case of library functions, we tend to do it all the time. For example, to set Y to the square root of X in Pascal, we unhesitatingly write

$$\textbf{Y} := \textbf{SQRT(X)}$$

We have a good sense of what the **SQRT** function does at a macroscopic level— it calculates square roots. At the same time, we have no clue as to how it does it. There are many possible algorithms for computing square roots. Some implementations may expand the appropriate Maclauren series; others might use Newton's method or even reduce the square root problem to its equivalent expression in terms of logarithms and exponentials. In any case, the user has little concern for the details.

As another example, if **N** is an integer and we use the Pascal statement

$$\textbf{WRITE(N : 1)}$$

Pascal will write out the value of **N** in a field which is exactly as large as it needs to be for that number. In programming, we use this statement freely, and are quite unconcerned about the details of that operation. In point of fact, this operation is considerably more complicated than it appears.

The reason that printing a number is complicated arises from the fact that most systems are not capable of printing numeric values directly. Instead, any number must first be converted into the appropriate sequence of characters, which can then be transmitted to the output device. For example, if N is 1492, the actual output will consist of the characters '1', '4', '9', and '2', in that order.

To make the problem specification concrete (and simplify it slightly at the same time), we will define the following as our goal:

In Pascal, write a procedure **PRINTNUM(N),** which takes a *nonneg-ative* integer N, and writes it out without any extraneous spaces.* As such, **PRINTNUM(N)** should simulate the function of

WRITE(N : 1)

but is allowed to perform only *character* output.

The complete solution to the problem consists of two largely independent tasks: (1) breaking up a large number into its component digits and (2) translating each digit into its character code. The second is somewhat easier, and it is useful to consider it first.

In Pascal, each value of type **CHAR** is represented internally by a numeric code. Given any character, we can determine the corresponding internal code through the use of the built-in function **ORD.** Similarly, given a numeric value within the legal character range, we can convert it into a character by using the **CHR** function. For example, on a system which uses ASCII (American Standard Code for Information Interchange),

ORD('A') = 65

and

CHR(65) = 'A'.

The exact correspondence between characters and numeric codes, how-ever, will differ from system to system, since it depends on which *character set* is used. Although a large percentage of modern implementations use ASCII, others may use EBCDIC (Extended Binary-Coded-Decimal Interchange Code) or even less common representational encodings. In each of these, the character values are different. Thus, programs should never assume that **ORD('A')** is 65, since that assumption may be incorrect on some systems.

Even though Pascal does not require the use of a specific character set, it does enforce certain restrictions on the internal representation. In particular, Pascal requires that the internal codes for the characters '0' through '9' be *consecutively increasing.* Thus, if the character '0' has the numeric value 48 (as it does in ASCII), then '1' must be represented as 49, '2' by 50, and so forth. In a different character set, '0' may correspond to some other value, but the remaining digits must follow in sequence. This requirement makes it pos-sible to write a Pascal function **DIGITCHAR,** which converts an integer in the range 0 through 9 into its corresponding character code:

---

*The constraint that N be nonnegative is included here only to simplify the **PRINTNUM** task. Removing this constraint is discussed in Exercise 4-2.

```
FUNCTION DIGITCHAR(D : INTEGER) : CHAR;
BEGIN
   DIGITCHAR := CHR(D + ORD('0'))
END;
```

Armed with this function, we can proceed to the problem of dividing a number into its component digits. Intuitively, we would like to design a loop structure which would select each digit of the number in turn, convert that digit to a character, and print it. If we follow the discipline of top-down design and leave the details to subroutines, we might start with the following definition:

```
PROCEDURE PRINTNUM(N : INTEGER);
   VAR
      I : INTEGER;
   BEGIN
      FOR I := 1 TO NDIGITS(N) DO
         WRITE(DIGITCHAR(ITHDIGIT(N, I)))
END;
```

Here, we assume that **NDIGITS(N)** returns the number of digits in a number **N** and that **ITHDIGIT(N, I)** returns the value of the Ith digit of **N**, numbering from the left. If we write these routines correctly, **PRINTNUM** will operate as advertised.

Unfortunately, neither of these routines is easy to code, particularly if efficiency is of any concern. In fact, after making a few attempts (see Exercise 4-1), it is reasonably likely that we would abandon this course and search for an alternative approach.

To determine if recursion can be helpful in this case, we must try to find a subdivision of the problem which results in a simpler instance of the same problem. For example, if we can find a way to transform the problem of printing a number into the problem of printing a smaller one, we will be well on the way to a recursive solution.

In searching for a recursive subdivision, it is not enough to find instances of simple subproblems within the context of a larger problem. It is equally important to find some way to reassemble the solutions to those simpler subproblems to generate a complete solution. For example, in writing **PRINTNUM(N)** recursively, it would certainly be legitimate to call **PRINTNUM(N-1)**. On the other hand, it is hardly helpful. If N has the value 2000, calling **PRINTNUM(N-1)** would print out '1999'. Clearly, this does not help much toward the goal of printing the number '2000'.

It would be more useful if there were some way to divide a number along digit boundaries so that the solution would consist of printing two strings side by side. The key to finding the appropriate subdivision lies in discovering that it is possible to split a number into its leading digits and its final digit by using the operations **DIV** and **MOD**. Given any integer **N,** the expression

$$N \text{ MOD } 10$$

always gives the value of the final digit, irrespective of the size of the original number. The digits which precede the final digit can be determined by the expression

$$N \text{ DIV } 10$$

For example, given **N** = 1492, **N DIV 10** is 149 and **N MOD 10** is 2. To print out the entire number, all we need to do is write out 149 and then write out the 2, with no intervening spaces.

This decomposition suggests the beginning of a recursive implementation. If we write the outline of the procedure in Pascal, but leave the internal operations in English, we are left with the following procedure "definition":

```
PROCEDURE PRINTNUM(N : INTEGER);
  BEGIN
    Print the number N DIV 10;
    Print the character corresponding to N MOD 10
  END;
```

This does not as yet constitute a complete recursive algorithm, since it provides only the decompositional rule and fails to define a simple case. Fortunately, defining an appropriate simple case is reasonably straightforward here. If a number has only one digit, we can simply convert the number to a character and print it out. Adding this to the informal definition gives us the following algorithmic specification:

```
PROCEDURE PRINTNUM(N : INTEGER);
  BEGIN
    IF N < 10 THEN
      Print the character corresponding to N
    ELSE
      BEGIN
        Print the number N DIV 10;
        Print the character corresponding to N MOD 10
      END
  END;
```

From here, all we need to do is replace the English commands with their Pascal equivalents. For two of those commands, this is reasonably simple since all the hard work has already been done in **DIGITCHAR**. After this refinement, the procedure has the form

```
PROCEDURE PRINTNUM(N : INTEGER);
  BEGIN
  IF N < 10 THEN
    WRITE(DIGITCHAR(N))
  ELSE
    BEGIN
    Print the number N DIV 10;
    WRITE(DIGITCHAR(N MOD 10))
    END
END;
```

Clearly, we are almost finished. All that remains is a single English phrase to be translated into Pascal:

"Print the number N DIV 10"

Fortunately, we are in the process of designing a procedure which does exactly that. Thus, we have at our disposal a new command which performs this operation, and we can write this using a recursive call:

PRINTNUM(N DIV 10)

Thus, the complete definition of **PRINTNUM** is as follows:

```
PROCEDURE PRINTNUM(N : INTEGER);
  BEGIN
  IF N < 10 THEN
    WRITE(DIGITCHAR(N))
  ELSE
    BEGIN
    PRINTNUM(N DIV 10);
    WRITE(DIGITCHAR(N MOD 10))
    END
END;
```

Taking the final step in the above derivation does not require cleverness so much as courage. The key to this example lies in a willingness to translate the informal algorithmic statement "print a number" into a call to the **PRINTNUM** procedure itself.

4-2   Generating a Primer (with apologies to Dick and Jane)

> This is Spot.
> See Spot run.
> Run, Spot, run.

For those who attended elementary school in the United States, sentences like this are almost certainly familiar. For many years, first-graders were taught to read through the use of primer texts designed to develop reading skills through the repetition of simple sentence patterns. Of these, the most widely circulated was the *Dick and Jane* series, in which we were introduced to the quintessentially suburban lives of Dick, Jane, and Sally, along with their pets Puff and Spot.

By their very nature, primers use a limited vocabulary and a highly restricted sentence structure. Presumably, this makes them more easily understood by those who are just learning to read. If we look at this from the opposite perspective, this same simplicity of structure should also make this sort of text easier to write. In particular, if we want to design a program that generates English text, *Dick and Jane* may prove to be ideal.

As a first step toward solving this problem, we should take advantage of the fact that many of the sentences in the typical *Dick and Jane* story are drawn from a rather simple set of patterns. For example, as the characters are introduced, we usually see a sentence of the following form:

$$\text{This is } \underline{\hspace{3cm}} \qquad \text{[Pattern \#1]}$$
$$\text{\textit{(name)}}$$

where the blank space is filled in with the appropriate name. We will refer to this as sentence pattern #1. In addition to pattern #1, there are several other patterns which occur quite frequently in *Dick and Jane*. For example, we encounter sentences having the following forms:

$$\text{See } \underline{\hspace{3cm}} \qquad \text{[Pattern \#2]}$$
$$\text{\textit{(name)}}$$

or

$$\text{See } \underline{\hspace{2.5cm}} \ \underline{\hspace{2.5cm}} \qquad \text{[Pattern \#3]}$$
$$\text{\textit{(name)}} \qquad \text{\textit{(verb)}}$$

The last pattern contains two blanks for substitution, and we can fill each one independently with a word of the appropriate class. For example, if our collection of names consists of "Dick," "Jane," "Sally," "Spot," and "Puff" and the available verbs are "run," "jump," and "play," we can create fifteen different sentences from pattern #3 alone.

Although we can add more patterns to this list, it is probably useful to stop at this point and turn our attention to the problem of generating random sentences in Pascal. The principles of top-down design suggest starting at the sentence level, but we get a better feeling for the necessary structure if we begin the implementation at a simpler level. For instance, our first step might be to write a procedure **NAME** which prints out the name of one of the five main characters. Ordinarily, this is accomplished by choosing a random integer in the range 1 to 5 and using that to select one of the five names.

Since we will often be making choices of this kind, it will be helpful to define a function **CHOOSE(N)** which returns a random integer in the 1 to N range. Unfortunately, the details of the **CHOOSE** function are rather difficult to specify because random number generation is not defined as part of standard Pascal. Nonetheless, most systems implement some mechanism for generating random numbers, even though it may differ from machine to machine. Typically, a system will provide a library function **RANDOM** which takes an arbitrary argument and returns a random real number, uniformly distributed in the interval (0.0, 1.0). To make this value correspond to our definition of **CHOOSE,** we need to scale this value to the proper range using a function such as:

```
FUNCTION CHOOSE(N : INTEGER) : INTEGER;
BEGIN
    CHOOSE := 1 + TRUNC(N * RANDOM(0))
END;
```

Even though most definitions of **CHOOSE** will be quite similar to this model, it is important to emphasize that this function is not *portable,* in the sense that changes may be necessary before it is used on a particular computer system.

Once this detail is out of the way, we can easily complete the coding of the **NAME** procedure:

```
PROCEDURE NAME;
BEGIN
    CASE CHOOSE(5) OF
        1 : WRITE('Dick');
        2 : WRITE('Jane');
        3 : WRITE('Sally');
        4 : WRITE('Spot');
        5 : WRITE('Puff')
    END
END;
```

We can use this same technique to generate the verbs:

```
PROCEDURE VERB;
BEGIN
CASE CHOOSE(3) OF
    1 : WRITE('run');
    2 : WRITE('jump');
    3 : WRITE('play')
END
END;
```

From here, we can easily define procedures which write out sentences according to each of the patterns in our current collection:

```
PROCEDURE PAT1;                    PROCEDURE PAT3;
BEGIN                              BEGIN
  WRITE('This is ');                WRITE('See ');
  NAME                             NAME;
END;                              WRITE(' ');
                                  VERB
PROCEDURE PAT2;                   END;
BEGIN
  WRITE('See ');
  NAME
END;
```

Finally, we can construct a random sentence-generating procedure simply by adding one more level to the structure:

```
PROCEDURE SENTENCE;
BEGIN
CASE CHOOSE(3) OF
    1 : PAT1;
    2 : PAT2;
    3 : PAT3
END;
WRITELN('.')
END;
```

So far, this application depends entirely on procedural decomposition to accomplish its goal. Starting from the sentence level, we first choose a pattern, break that down into its components, and then proceed to select random words as necessary to complete the pattern. At each step in the process, we adhere to the discipline of considering only a single procedural level at a time. For example, looking at the code

```
PROCEDURE PAT3;
BEGIN
  WRITE('See ');
  NAME;
  WRITE(' ');
  VERB
END;
```

tells us everything we need to know about the pattern at this level; the details of choosing a noun and verb are deferred to the appropriate procedures.

As of now, the sentence-generating program is still a bit on the boring side, since it only handles proper names and a few simple verbs. To make things a little more exciting, we can extend the program by adding a new sentence pattern:

_____ _____ _____   [Pattern #4]
  (name)   (transitive verb) (noun phrase)

The idea here is to generate sentences of the form

Jane hit the red ball.

or

Spot found the little green stick.

Conceptually, most of this is quite straightforward. First, we need to define a **PAT4** procedure and add a corresponding line to **SENTENCE** so that **PAT4** will be included in the list of available patterns.

```
PROCEDURE PAT4;
BEGIN
  NAME;
  WRITE(' ');
  TRANSVERB;
  WRITE(' ');
  NOUNPHRASE
END;
```

For use with this pattern, we can also define a few transitive verbs:*

---

*Note that it is not possible to define a single verb category and share that between patterns #3 and #4, since the form of the verb is different in the two patterns.

```
PROCEDURE TRANSVERB;
BEGIN
  CASE CHOOSE(4) OF
    1 : WRITE('watched');
    2 : WRITE('liked');
    3 : WRITE('found');
    4 : WRITE('hit')
  END
END;
```

At this point, we are left with the problem of generating a "noun phrase." From the sample sentences given above, a noun phrase consists of the word "the," possibly followed by one or more adjectives, and finally a noun. Just to get all the preliminaries out of the way, we can define the procedures NOUN and ADJECTIVE to generate words which fit that category in the pattern:

```
PROCEDURE NOUN;                    PROCEDURE ADJECTIVE;
BEGIN                              BEGIN
  CASE CHOOSE(3) OF                  CASE CHOOSE(4) OF
    1 : WRITE('ball');                 1 : WRITE('big');
    2 : WRITE('stick');                2 : WRITE('little');
    3 : WRITE('house')                 3 : WRITE('red');
  END                                  4 : WRITE('green')
END;                                 END
                                   END;
```

From here, we must finally turn to the problem of the "noun phrase" itself and determine how the corresponding procedure NOUNPHRASE might best be defined.

Internally, the structure of a noun phrase is somewhat more complex than the examples considered earlier, since it contains an element (specifically, the adjective) which can be repeated within the pattern. If we think about this repeated operation as a FOR loop, we come up with an iterative procedure with the following form:

```
PROCEDURE NOUNPHRASE;
VAR
  I : INTEGER;
BEGIN
  WRITE('the ');
  FOR I := 1 TO CHOOSE(3) DO
    BEGIN
      ADJECTIVE;
      WRITE(' ')
    END;
  NOUN
END;
```

This procedure approximates the desired effect, but there are two rather important problems. Although we could change the limits in the **FOR** loop so that the program could produce more than three adjectives, the maximum number of adjectives is always limited by a fixed upper bound. While this may be a useful limitation in the context of a primer text, the corresponding pattern in English allows any number (including zero) of adjectives to modify the noun. The second problem is that **NOUNPHRASE** has an entirely different structure from each of the other procedures we have defined, in that it contains a **FOR** loop along with local variables. If this program is to be easily extended and modified, it is helpful to preserve as symmetrical a structure as possible.

We can find a solution to these problems by thinking about the structure of a noun phrase in a slightly different way. At present, we have a conceptual pattern for a noun phrase which can be diagrammed as follows:

> the _____ _____
>   *(several adjectives)*   *(noun)*

As an alternative, we can define the noun-phrase pattern using a two-level structure. To start with, we can separate out the problem of the word "the" by defining a noun phrase as having the following pattern:

> the _____
>   *(modified noun)*

Moving one step further, a modified noun consists of one of two possibilities: (1) a simple noun or (2) an adjective *followed by a modified noun*. For example, "house" is a legal modified noun because it conforms to the first possibility. So is "red house," because "red" is an adjective and "house" is a modified noun. Similarly, "big red house" is a modified noun, since it also consists of an adjective and a modified noun.

Since this approach defines a modified noun in terms of other modified nouns, the definition is itself recursive, and it would not be surprising to see recursion used in the implementation. In fact, once the definition is out of the way, the procedures used for a noun phrase are quite straightforward, since they again correspond closely to the conceptual definition:

```
PROCEDURE MODNOUN;
BEGIN
  CASE CHOOSE(2) OF
    1 : NOUN;
    2 : BEGIN
          ADJECTIVE;
          WRITE(' ');
          MODNOUN
        END
  END
END;
```

```
PROCEDURE NOUNPHRASE;
BEGIN
  WRITE('the ');
  MODNOUN
END;
```

In this chapter, we have produced a relatively powerful primer generator, which can produce the following sentences:

> This is Spot.
> See Spot run.
> Sally watched the ball.
> Jane liked the little house.
> Spot found the big red stick.

These, of course, seem perfectly in character. Unfortunately, the program can equally well come up with sentences of questionable literary merit, such as

> Dick watched the little big house.

or even such obvious monstrosities as

> Spot liked the green little red red big red ball.

The problem here is that the program is concerned only with the form or *syntax* of its output and takes no account of its content.

The technique used to generate the sentences in this section is based on a linguistic structure called a *context-free grammar*. The name is derived from the fact that, within any pattern, the legal substitutions are not restricted by the surrounding context, even if this results in meaningless combinations. Although this approach has some obvious drawbacks when used to generate English text,* context-free grammars are extremely important to the translation of programming languages and will be covered in detail by more advanced courses.

---

*On the other hand, the use of context-free grammars has hidden virtues. Since all names are chosen without regard to context, the generated sentences are much less likely to reflect cultural biases and stereotypes. Thus, it is just as likely to generate "Jane hit the ball" or "Dick liked the little house" as the other way around.

Bibliographic Notes

When textbooks give examples of recursive procedures, they are usually at a more difficult level than the ones presented in this chapter. Some of the best sources for additional exercises at this level are found in Ford [1982 and 1984]. Certain exercises from Tenenbaum and Augenstein [1981] are appropriate as well.

Exercises

4-1. Without using recursion, complete the iterative coding of **PRINTNUM** by defining the functions **NDIGITS** and **ITHDIGIT.** In terms of the number of digits in the input number, what is the complexity order of your algorithm?

4-2. As written, **PRINTNUM** does not handle negative numbers. We can fix this within the recursive coding as follows:

```
PROCEDURE PRINTNUM(N : INTEGER);
BEGIN
  IF N < 0 THEN
    BEGIN
      WRITE('-');
      PRINTNUM(-N)
    END
  ELSE
    IF N < 10 THEN
      WRITE(DIGITCHAR(N))
    ELSE
      BEGIN
        PRINTNUM(N DIV 10);
        WRITE(DIGITCHAR(N MOD 10))
      END
END;
```

What are the disadvantages of this particular coding? How might you correct this?

4-3. In several of the applications that we encounter later in this book, we will see that zero often makes a better simple case than one. For example, in the **PRINTNUM** problem, we used a one-digit number as the simple case. However, printing a number with no digits is even easier, and we might be tempted to write **PRINTNUM** as follows:

```
PROCEDURE PRINTNUM(N : INTEGER);
  BEGIN
   IF N > 0 THEN
    BEGIN
     PRINTNUM(N DIV 10);
     WRITE(DIGITCHAR(N MOD 10))
    END
  END;
```

This program, however, does not always produce exactly the same output as the version in the text. Describe the difference.

4-4. Using the recursive decomposition from **PRINTNUM,** write a function **DIGITSUM** which takes an integer and returns the sum of all its digits.

4-5. The *digital root* of a number is calculated by taking the sum of all the digits in a number and then repeating that process with the resulting sum until only a single digit remains. For example, if we start with 1969, we first add the digits to get 25. Since this is more than a single digit, we must repeat the operation, giving 7 as the final answer.

Starting with the **DIGITSUM** function from exercise 4-4, write a function **DIGITALROOT(N)** which calculates this value.

4-6. Extend the definition of a "noun phrase" so that proper names are also acceptable. This would allow sentences such as

Sally liked Jane.

or

Spot found Dick.

4-7. As written, the most general routine in the *Dick and Jane* example is the **SENTENCE** procedure, which generates a single sentence using a random pattern. Extend this by adding a **PARAGRAPH** procedure, which generates a "paragraph" consisting of a random number of sentences. Make sure that the **PARAGRAPH** procedure operates recursively and uses the same general structure as the other procedures designed for this chapter.

4-8. Following the general outline of the primer-generation example, write a collection of procedures that will generate syntactically correct (albeit meaningless) Pascal programs. This program should not try to incorporate all of Pascal's structure, but it should include many of the more common forms.

The Tower of Hanoi

"Firstly, I would like to move this pile from here to there," he explained,
pointing to an enormous mound of fine sand; "but I'm afraid that all I
have is this tiny tweezers." And he gave them to Milo, who immediately
began transporting one grain at a time.
                              —*Norton Juster,* The Phantom Tollbooth

Toward the end of the nineteenth century, a new puzzle appeared in Europe
that quickly became quite popular on the continent. In part, its success can be
attributed to the legend that accompanied the puzzle, which was recorded in
La Nature by the French mathematician De Parville (as translated by the
mathematical historian W. W. R. Ball):

> In the great temple at Benares beneath the dome which marks the
> center of the world, rests a brass plate in which are fixed three diamond
> needles, each a cubit high and as thick as the body of a bee. On one
> of these needles, at the creation, God placed sixty-four disks of pure
> gold, the largest disk resting on the brass plate and the others getting
> smaller and smaller up to the top one. This is the Tower of Brahma.
> Day and night unceasingly, the priests transfer the disks from one
> diamond needle to another according to the fixed and immutable laws
> of Brahma, which require that the priest on duty must not move more
> than one disk at a time and that he must place this disk on a needle
> so that there is no smaller disk below it. When all the sixty-four disks
> shall have been thus transferred from the needle on which at the
> creation God placed them to one of the other needles, tower, temple
> and Brahmins alike will crumble into dust, and with a thunderclap the
> world will vanish.

As is so often the case with legend, this tale has evolved in the telling,
even to the point of acquiring a new geographic setting. In more modern times,
the Tower of Brahma has become known as the Tower of Hanoi, and it is in
this form that it has been passed on to today's students as the classic example
of procedural recursion.

Before attempting to tackle the complete problem with 64 golden disks
(gold prices being what they are), it is useful to consider a simplified version

of the puzzle. For example, if there are only six disks, the initial state of the puzzle can be represented with all the disks on needle A:

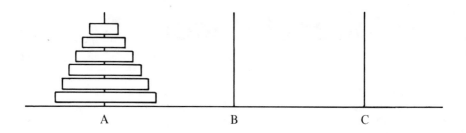

The object is to perform the right sequence of moves so that all the disks end up on needle B:

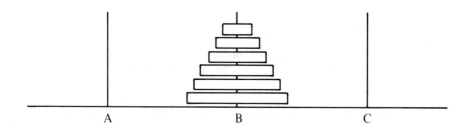

Of course, by the laws of the game (and to avoid ending the universe prematurely), we are not allowed to pick up the whole stack and move it as a unit. We are therefore faced with the problem of breaking down this high-level solution into a series of individual moves.

5-1   The Recursive Solution

Recursively, the critical point to notice is that the complete goal (in this case, moving the entire stack of six disks from needle A to needle B) can be broken down into the following subgoals:

    1. Move the top five disks from A to C:

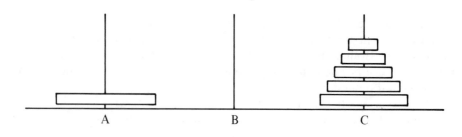

2. Move the bottom disk from A to B:

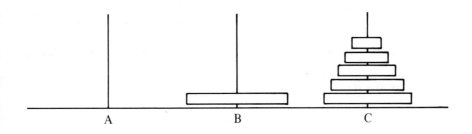

3. Move the top five disks back from C to B, leaving the desired final state:

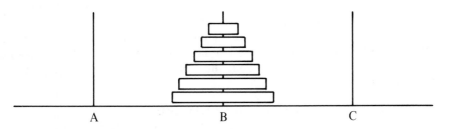

Once again, given the rules of the game, we cannot move the entire set of five disks as a unit. On the other hand, there is certainly no rule that prohibits *thinking* about a five-disk transfer as a single operation, and using that as a way of simplifying the problem. Thinking recursively, we know that if we can transform the six-disk problem into one that requires transferring a stack containing only five disks, we are well on the way to a complete solution.

We do, however, need to exercise some caution here. In breaking the problem down into a simpler case, we must be certain that the new subproblem has precisely the same form. Here, for example, it is important to verify that we can apply the recursive technique to the tower of five disks without violating any of the rules of the games. In particular, we must ensure that a further decomposition of this task will not force us to move a larger disk on top of a smaller one.

Given this recursive decomposition, this does not turn out to be a problem. Whenever we move a tower as part of a recursive operation, all the disks in that tower are *smaller* than any of the other disks that might already be on one of the needles. Since they are smaller, we can freely move these disks from one tower to another, even if larger disks are already there.

Before we write a program to solve this puzzle, it helps to define the procedures involved in a relatively informal way. Here, what we need is a procedure which writes out all the moves necessary to transfer a tower of a given size. Since the operation involved is moving a tower, **MOVETOWER** seems like an appropriate name.

In designing a recursive procedure like **MOVETOWER**, it is essential to remember that the procedure is not only responsible for the complete problem as perceived by the outside user, but for a variety of internally generated subproblems as well. Thus, although it is tempting to think of **MOVETOWER** as having only one argument indicating the number of disks, this will not provide enough information to solve the general case. The statement

MOVETOWER(6)

tells us the number of disks involved but gives us no insight as to where they start or where they finish. Even though the complete problem might always move disks from A to B, the recursive calls will not. For example, somewhere deep in the recursive structure the program will need to be able to "move a tower of size 3 from C to A using B for temporary storage."

In order to make **MOVETOWER** sufficiently general, the parameters to **MOVETOWER** must include which needles are involved, in addition to the number of disks. Thus, we need the following procedure header for **MOVETOWER:**

PROCEDURE MOVETOWER(N:INTEGER;
START, FINISH, TEMP:CHAR);

where N is the number of disks to move and **START, FINISH** and **TEMP** indicate the role currently assigned to each needle in the current subproblem. Thus, in the main program to solve the six-disk problem, we would expect the call

MOVETOWER(6, 'A', 'B', 'C')

which can be rendered in English as the sentence "move a tower of size 6 from A to B using C as the temporary needle." Internally, if we do reach a position in the algorithm where we need to "move a tower of size 3 from C to A using B for temporary storage," that would be represented as a call with the values

MOVETOWER(3, 'C', 'A', 'B')

The complete solution also requires that we find a simple case which allows the recursion to terminate. Finding the simple case is usually much more straightforward than determining the recursive decomposition, and this is indeed true in the Tower of Hanoi problem. Although we will consider another possibility in Exercise 5-2, an obvious choice is the case when there is only a single disk to move. This leads to a recursive solution with the following form:

If N (the number of disks to move) is one, simply move that disk from **START** to **FINISH.**

If N > 1, divide the problem up into three subgoals:

1. Using this same algorithm, move the top N–1 disks from **START** to **TEMP.** In the process of making this transfer, the **FINISH** needle will be used as the new temporary repository.
2. Move the bottom disk from **START** to **FINISH.**
3. Move the top N–1 disks back from **TEMP** to **FINISH,** using **START** as the new temporary.

Given this informal algorithmic description, we can proceed to the problem of coding it in Pascal. As in the case of most recursive routines, the procedure definition corresponds quite closely to the algorithmic definition:

```
PROCEDURE MOVETOWER(N : INTEGER;
                    START, FINISH, TEMP : CHAR);
BEGIN
  IF N = 1 THEN
    WRITELN(START, ' to ', FINISH)
  ELSE
    BEGIN
      MOVETOWER(N–1, START, TEMP, FINISH);
      WRITELN(START, ' to ', FINISH);
      MOVETOWER(N–1, TEMP, FINISH, START)
    END
END;
```

If this procedure is compiled together with a main program consisting of the single statement

MOVETOWER(6, 'A', 'B', 'C')

the program will write out a complete history of the moves necessary to solve the six-disk Hanoi problem.

But how? This is the question that always seems to arise whenever the **MOVETOWER** procedure is presented. The recursive decomposition is all well and good as a theoretical idea, but this is a program definition! Where are the details? Where is the solution?

5-2   The Reductionistic View

One of the first things that new programmers learn is that computer programming requires a meticulous attitude with respect to details. When programming is involved, the statements "it's right except for a few details" and "it's wrong" are hard to differentiate experimentally. In a recursive implementation, those details are sometimes difficult to see, because the work consists primarily of keeping track of a list of active subproblems—details which are handled automatically for the programmer by the compiler.

To understand the underlying process behind that record keeping, it is useful to examine the execution history of a recursive program in complete detail. The Tower of Hanoi problem has two characteristics which make it particularly appropriate for this analysis. The problem is (1) hard enough that the solution is not obvious and (2) simple enough that we can follow the recursive logic without getting lost. With any luck, going through the details here will dispel any lingering reductionistic tendencies and provide additional confidence in the recursive approach.

To simplify the basic problem, we will consider the necessary operations when only three disks are involved. Thus, the main program call is

MOVETOWER(3, 'A', 'B', 'C')

From here, all we need to do is keep track of the operation of the program, particularly as it makes new calls to the **MOVETOWER** procedure.

Following the logic of a recursive procedure is often tricky and may not be advisable for the fainthearted. The best way, of course, is to let the computer handle all of this, but our purpose here is to expose those details. To do this, we must "play computer" and simulate its internal operations. For recursive problems, one of the best ways of keeping track of all the necessary operations is with a stack of 3×5 index cards, where a new card is used every time a new procedure call is made.

Since we have just made the procedure call

MOVETOWER(3, 'A', 'B', 'C')

we are ready for our first index card. On each card, the first thing we do is write down the English interpretation of the call and refer to that as our current goal. Following that, we also specify the values for each parameter:

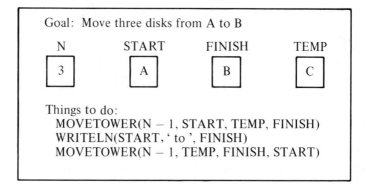

On the bottom of the card, we make a list of the various tasks we must perform to solve the problem *at this level*. Here, the "things to do" list is simply the statements that form the **ELSE** clause in the **MOVETOWER** procedure.

From here, we go through each of the subtasks in turn. Thus, in order to begin the solution at this level, we need to execute the statement

MOVETOWER(N–1, START, TEMP, FINISH)

As with any procedure call, the first step is the evaluation of the arguments. To do this, we need to find the values of the variables N, **START, TEMP** and **FINISH**. Whenever we need to find the value of a variable, we must use the value as it is defined *on the current index card*. Thus, the **MOVETOWER** call is equivalent to

MOVETOWER(2, 'A', 'C', 'B')

This operation, however, indicates a recursive call. This means that we must suspend our current operation and repeat this entire process with a new index card. On this new card, we simply copy in the parameter values in the order in which they appeared in the procedure call:

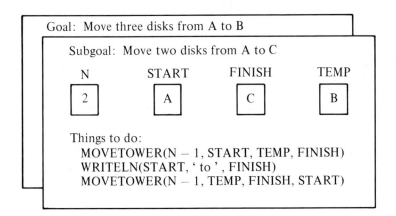

Our original index card still has several unfinished tasks, so we cannot dispense with it, but must set it aside until we complete the new subgoal. Thus, in the diagram above, the new card is placed on top of the previous one, hiding the previous values of the variables. As long as this index card is active, N will have the value 2, **START** will be 'A', **FINISH** will be 'C', and **TEMP** will be 'B'. When the operations on the new "things to do" list are complete, we will throw this card away, restoring the original card to the active position. In programming terminology, the information stored on each index card is called an *environment*.

Since we have just made a new call to the **MOVETOWER** routine, we must start over at the beginning of the procedure body. Once again, N is not 1, so that our first operation is

MOVETOWER(N–1, START, TEMP, FINISH)

Here, however, the parameters have changed. Substituting the new values (that is, the values on the current index card) into this statement gives

MOVETOWER(1, 'A', 'B', 'C')

Once again, we have a recursive call, giving rise to a third level of subgoal:

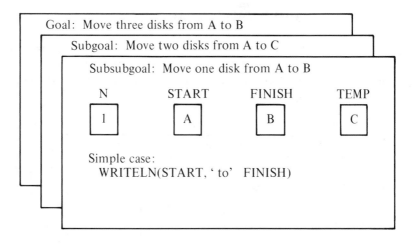

This, however, represents an easy case. All we need do is write out the message "A to B", and we are finished. This leaves us in the configuration:

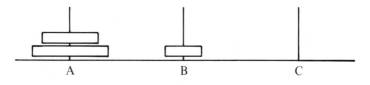

At this point, we are done with the current card and can remove it from our active list. This brings us back to the previous card, having just completed the first item on the "to-do" list. To indicate that this operation is complete, we simply cross it off and move on to the next item.

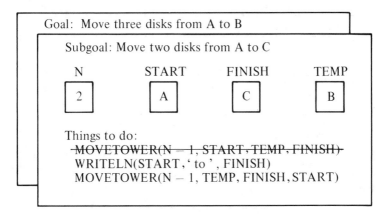

The next operation does not require any further calls; we simply move the indicated disk to reach the position:

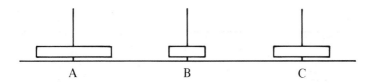

Since we have finished yet another operation on the current card, we can cross this off as well, leaving our current list of pending operations as follows:

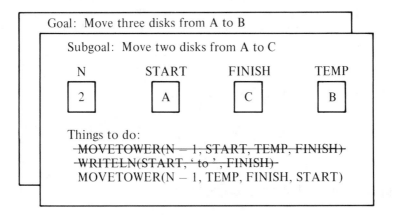

Here, we again need to "move a tower of size 1." Internally, we know that this will be handled by a new call to **MOVETOWER** (and will thus require a new index card), but we should be able to see by this point that the program will be able to handle this as a simple case and end up in the following state:

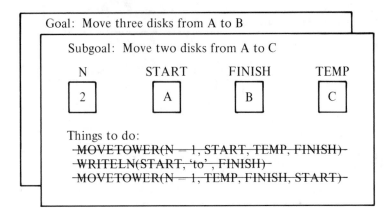

Once again, we are completely done with one of the index cards and can throw it away. This restores the first index card to our current operation pile with the first subtask completed.

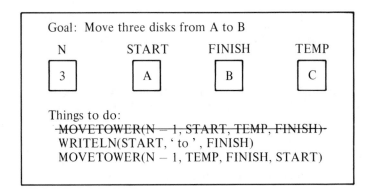

At this point, we have made three moves:

A to B
A to C
B to C

This puts us in the following position:

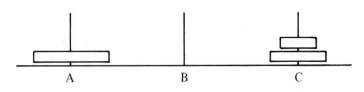

As this diagram illustrates, we have just completed the first major subgoal and are pretty much halfway through the process. From here, we simply move the largest disk from A to B and tackle the remaining subgoal. Again, this will require further decomposition, and you are encouraged to continue the analysis on your own.

Bibliographic Notes

The Tower of Hanoi is such a perfect example of recursion that it is used in almost all texts that include the topic. A thorough trace of the recursive operation at a more machine-oriented level is included in Tenenbaum and Augenstein [1981]. For a more popular account, Dewdney's article in *Scientific American* [1984] is quite good and includes a discussion of the more advanced "Chinese-ring" puzzle. This puzzle is also included in Ford [1984], along with other problems that are appropriate at this level.

Exercises

5-1. In making a recursive subdivision of a problem, it is extremely important to ensure that any generated subproblems obey exactly the same rules as the original. With this in mind, consider the following decomposition:

To move a tower of N disks:

1. Move the top disk from **START** to **TEMP**.
2. Using a recursive call, move the remaining tower of N–1 disks from **START** to **FINISH**.
3. Move the top disk back from **TEMP** to **FINISH**.

Why does this algorithm fail?

5-2. Although the version presented here is somewhat easier to explain, it is possible to write a simpler version of **MOVETOWER** by changing the simple case. Although the one-disk case is certainly simple, there is even less work required when there are no disks at all. Using this as the simple case gives **MOVETOWER** the form shown at the top of the next page.

```
PROCEDURE MOVETOWER(N : INTEGER;
                       START,FINISH,TEMP : CHAR);
 BEGIN
  IF N > 0 THEN
    BEGIN
      MOVETOWER(N-1, START, TEMP, FINISH);
      WRITELN(START, ' to ', FINISH);
      MOVETOWER(N-1, TEMP, FINISH, START)
    END
 END;
```

Although this is more concise than the original version, it is slightly less efficient. Starting with the statement

$$\text{MOVETOWER(3, 'A', 'B', 'C')}$$

how many calls to **MOVETOWER** are required to complete the transfer? How does this compare with the implementation used in the text?

5-3.   Which of you, intending to build a tower,
          sitteth not down first, and counteth the cost?
                              —Luke 14:28

Using mathematical induction, prove that the number of moves required to transfer a tower of size N by the **MOVETOWER** algorithm is given by

$$2^N - 1$$

Permutations

The order is rapidly fading
And the first one now will later be last
                    —*Bob Dylan, "The Times They Are A-Changin'"*

As part of the discussion of factorials in section 3-2, we introduced the term *permutation* to refer to an arrangement of objects in a linear order. At the time, our goal was simply to determine the number of possible permutations given a collection of N distinct objects. We reasoned that there are N ways to choose the first object, N–1 ways to choose the second, and so forth. This means that the total number of arrangements can be determined by calculating N!:

$$N \times (N-1) \times (N-2) \times \ldots \times 1$$

For example, given the string 'ABCD', there are twenty-four (4!) different permutations:

ABCD	BACD	CABD	DABC
ABDC	BADC	CADB	DACB
ACBD	BCAD	CBAD	DBAC
ACDB	BCDA	CBDA	DBCA
ADBC	BDAC	CDAB	DCAB
ADCB	BDCA	CDBA	DCBA

In this chapter, our goal is to write a program which will generate this list. Before we write the actual code, however, it is important to define the data structures to be used. In Pascal, strings are conventionally represented as a packed array of characters of a constant length. Thus, to define the type **STRING** suitable for storing 'ABCD', we will use the following **CONST** and **TYPE** declarations*:

---

*In practice (see Exercise 6-1), it is probably better to define the type **STRING** to be a much larger array and use a separate variable to keep track of the number of "active" characters. This would allow the same **PERMUTE** procedure to be used with strings of different length. By defining N as a constant in this example, we can simplify the problem somewhat and focus attention more directly on the recursive structure of the algorithm.

```
CONST
  N = 4;

TYPE
  STRING = PACKED ARRAY [1..N] OF CHAR;
```

Given this representation, we can turn our attention to the details of the Pascal implementation. From an "exterior" point of view (such as that of the main program), we would like to define a procedure **PERMUTE** so that calling

PERMUTE('ABCD')

corresponds to the English operation

Generate and display all possible permutations of 'ABCD'.

This provides an informal definition of the **PERMUTE** procedure and establishes its argument structure:

PROCEDURE PERMUTE(S : STRING);

In demonstrating that a set of **N** objects has **N** factorial permutations, we gave the basic outline of the recursive procedure necessary to generate those arrangements. Starting with the string 'ABCD', for example, there are four different ways to choose the first letter. This divides the possible permutations into four distinct groups:

> All permutations starting with 'A'
> All permutations starting with 'B'
> All permutations starting with 'C'
> All permutations starting with 'D'

If we consider any of these individual subproblems, however, we make an important discovery. The set of all permutations starting with 'A' is simply the permutations of 'ABCD' in which the 'A' is left alone and only the substring 'BCD' is affected by the recursive call. Similarly, to find the permutations beginning with 'B', we move the 'B' to the front and generate all permutations of 'BACD', again limiting any rearrangement to the last three characters. Thus, the entire process of generating all permutations of 'ABCD' can be decomposed into the following subgoals:

1. Generate the permutations of 'ABCD', changing only 'BCD'.
2. Generate the permutations of 'BACD', changing only 'ACD'.
3. Generate the permutations of 'CABD', changing only 'ABD'.
4. Generate the permutations of 'DABC', changing only 'ABC'.

This insight, however, forces us to change our conception of the problem. From the subproblem level, the principal operation is no longer

Generate and print all permutations of **S.**

but the somewhat more detailed statement

Generate and print all permutations of **S**, manipulating only those characters which occur between some position **K** and the end of the string.

To accommodate this change in the informal specification, we must also change the procedure header for **PERMUTE**. In addition to the string **S**, **PERMUTE** must also know the position which marks the boundary between the fixed letters and the letters which may still be manipulated by the recursive call. Thus, we must add a new argument **K** which indicates the first position in **S** which may be changed by this level of call:

PROCEDURE PERMUTE(S : STRING; K : INTEGER);

Thus, when the main program makes the call

PERMUTE('ABCD', 1)

this can be interpreted as "generate and print all permutations of 'ABCD', changing only the characters in positions 1 to 4." Since **K = 1** here, this indicates that the entire string is subject to change. As we proceed through the recursive process, however, **K** will increase by one at each level of call. As **K** increases, the number of characters which can be manipulated by this algorithm gets smaller, so that the problem becomes simpler at each level.

To ensure proper termination, we must define a simple case. Here, the case **K = N** seems most appropriate, as this implies that there is only one letter left. Since there is only one way to arrange a single letter, we can simply write out the current value of **S** and consider the problem solved.

We are now in a position to outline the general structure of the **PERMUTE** algorithm:

If **K = N**, write out **S** and return.
Otherwise, for each character in positions **K** through **N**:

1. Move that character to position **K** by exchanging it with the character formerly in that position.
2. Generate all permutations of **S**, working only with the positions **K + 1** through **N**.

The operation of exchanging two characters in a string is simple enough that we could perform it directly within the **PERMUTE** procedure. On the other hand, using a separate procedure for this purpose improves the overall program structure and makes the operation of the **PERMUTE** program considerably more clear. Thus, we will define a procedure **SWAPCHAR(S, P1, P2)** which exchanges the characters in **S** at positions **P1** and **P2**:

```
PROCEDURE SWAPCHAR(VAR S : STRING; P1, P2 : INTEGER);
  VAR
    TEMP : CHAR;
  BEGIN
    TEMP := S[P1];
    S[P1] := S[P2];
    S[P2] := TEMP
  END;
```

This procedure clears away the last hurdle to completing the definition of **PERMUTE**:

```
PROCEDURE PERMUTE(S : STRING; K : INTEGER);
  VAR
    I : INTEGER;
  BEGIN
    IF K = N THEN
      WRITELN(S)
    ELSE
    FOR I := K TO N DO
      BEGIN
        SWAPCHAR(S, K, I);
        PERMUTE(S, K + 1)
      END
  END;
```

Throughout the discussion of both the algorithm and its implementation, we have relied on an assumption which deserves more explicit attention. As we go through the **FOR** loop in the above procedure, the **SWAPCHAR** operation continually changes the contents of **S**. Assuming that **S** starts out as 'ABCD', the first cycle simply exchanges the 'A' with itself and therefore has no effect. When **I** = 2, the first and second characters of **S** are exchanged, so that **S** has the value 'BACD'. When **I** = 3, the first and third characters switch positions, resulting in 'CABD'.

These observations, however, are only true if the recursive **PERMUTE** call has no effect on **S**. Here, we have avoided the issue entirely by making **S** a *value parameter* to the **PERMUTE** procedure. At each level, the **PERMUTE** procedure receives a *copy* of the string in the next higher level. Within that level, any changes made to **S** are made only to the copy and therefore cannot affect the original argument.

Bibliographic Notes

There are several distinct approaches to the problem of generating permutations, and each textbook chooses its own conceptual model to present this problem. Knuth [1973] presents several methods for generating permutations and includes some historical background on the use of permutations in mathematics. Generating permutations is also discussed in Tenenbaum and Augenstein [1981] and Sedgewick [1983].

Exercises

6-1. In this program, we can improve on the data structure design by declaring **STRING** as a much larger array:

```
CONST
  MAXSTR = 20;

TYPE
  STRING = PACKED ARRAY [1..MAXSTR] OF CHAR;
```

Whenever a procedure requires a string-valued argument, we will pass *two* parameters: (1) a variable of type **STRING** which contains the characters and (2) an **INTEGER** which specifies the number of characters actually in use.

Change the implementation of **PERMUTE** so that it accepts a string of any length (up to **MAXSTR**) and prints out a list of its permutations. In testing this program, remember that a string of length 5 already has 120 permutations, so the sample runs should probably be kept small.

6-2. Using the index-card method described in Chapter 5, go through the operation of the **PERMUTE** program for the string 'ABCD'. Remember that each call to **PERMUTE** has its own copy of the string **S**. Thus, any changes which are made to **S** by **SWAPCHAR** are *local* to that level and are thus made only on the current index card.

6-3. Using mathematical induction, prove the following proposition:

If the characters in **S** are in alphabetical order, as in 'ABCD', **PERMUTE(S)** will generate the permutations in alphabetical order.

6-4. Throughout the discussion of permutations, we have required each letter to be distinct. The reason for this restriction is that the algorithm given will generate multiple copies of the same arrangement if any letter appears more than once. For example, if we attempt to list the permutations of 'ABBC' using this algorithm, each entry will appear twice.

Consider the following suggested implementation:

```
PROCEDURE PERMUTE(S : STRING; K : INTEGER);
  VAR
    I : INTEGER;
  BEGIN
    IF K = N THEN
      WRITELN(S)
    ELSE
      FOR I := K TO N DO
        IF S[K] <> S[I] THEN
          BEGIN
            SWAPCHAR(S, K, I);
            PERMUTE(S, K+1)
          END
  END;
```

This is the same procedure as before, except for the addition of the **IF** statement within the **FOR** loop. The intent of this statement was to eliminate any permutation which had been generated before by checking for the duplicate letter.

Unfortunately, this implementation is seriously flawed. What does it print out given the string 'ABBD'? What logical errors have been made in its design?

Rewrite this program so that it operates correctly when given a string with repeated letters. For example, given 'ABBD', the program should produce

ABBD	BABD	BBDA	DABB
ABDB	BADB	BDAB	DBAB
ADBB	BBAD	BDBA	DBBA

6-5. Like permutations, the problem of generating all subsets of a set of **N** elements has a simple recursive formulation. For example, if we start with the set {A,B,C}, the set of all subsets can be divided into two groups: (1) those that contain A and (2) those that do not. In either case, the subsets in each group simply contain all possible subsets of the remaining elements {B,C}. Thus, the complete list of subsets of {A,B,C} is

Group 1:	{A,B,C}	{A,B}	{A,C}	{A}
Group 2:	{B,C}	{B}	{C}	{ }

Using a string of characters to represent a set, write a recursive program which generates all subsets of {A,B,C,D}.

6-6. Rewrite Exercise 6-5 using Pascal's **SET** type instead of strings to represent the set of characters.

6-7. On a standard telephone, each of the digits (except 1 and 0) is associated with three letters as follows:

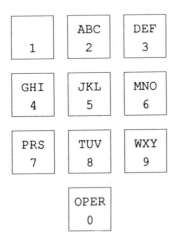

For certain telephone numbers, this makes it possible to find words which correspond to the necessary digits. For example, in the Boston area, a recorded message giving the time of day is available by dialing "NER-VOUS" (637-8687).

Write a program which accepts a string of digits and prints out all sequences of letters which correspond to that telephone number.

Sorting

In business and commercial applications, programming typically consists of a relatively simple set of operations. For example, report generation is often limited to reading in a file of data and printing it out again in some specific format, possibly including some summary calculations along the way. As long as the operations remain this easy to perform, the fact that these applications also tend to work with vast quantities of data is not an insurmountable concern. Operations such as formatting a report or adding up a column of figures tend to run in linear time. As discussed in Chapter 2, this implies that doubling the size of the data merely doubles the amount of time required. For most applications, this increase in running time seems entirely reasonable and well within acceptable bounds.

Unfortunately, not all the operations that characterize business-oriented computing are quite so well-behaved. Many applications that work with large amounts of data require that data to be arranged in some sequential order. Membership lists and telephone directories, for example, are arranged alphabetically to make each individual entry easier to find. Similarly, bulk mailings are arranged according to ZIP codes to meet the requirements of the U.S. Postal Service. The process of taking an unordered set of data and arranging it in sequence is called *sorting*.

Unlike many other operations that arise frequently in practice, sorting cannot be performed in linear time. When sorting is involved, doubling the number of items to be sorted will require more than twice as much time. As the size of the data grows large, this can result in a tremendous explosion in the required computation time. To control this phenomenon, it is essential to choose an algorithm which reduces the computational complexity as much as possible.

In this chapter, we will consider two distinct algorithms for sorting and compare their behavior. The first is called *selection sorting*, which operates iteratively by selecting the smallest element on the first cycle of a loop, followed by the second smallest, and so forth. The second method is called *merge sorting* and is characterized by the recursive process of (1) dividing an array in half, (2) sorting each piece, and (3) reassembling the sorted subarrays. Conceptually, selection sorting is quite intuitive, while merge sorting seems complex. However, as the number of items to be sorted becomes large, merge sorting turns out to be considerably more efficient.

In describing each of these algorithms, we will simplify the problem by assuming that we are sorting an array of integers. Moreover, we will also make the illustrations more concrete by naming the array **A** and assuming it is initialized to contain the following eight values:

A

90	99	63	82	93	76	81	76
1	2	3	4	5	6	7	8

Our goal is to write a procedure **SORT** which rearranges these numbers into increasing order.

To make the structure of the implementation more specific, we can write out an explicit procedure header for **SORT**, even though we have not yet written the code. Here, **SORT** needs access to the array of integers **A**. Since the elements of that array will be changed during the operation of **SORT**, the array must be passed as a **VAR** parameter. Moreover, since it is often necessary in Pascal to declare a large array and use only part of it, we should also pass the number of elements to be considered. This makes it possible to use **SORT** on arrays which differ in their number of active elements. Thus, the procedure header for **SORT** should be

PROCEDURE SORT(VAR A : INTARRAY; N : INTEGER);

where **INTARRAY** is defined in the **TYPE** section as

INTARRAY = ARRAY [1..MAXSIZE] OF INTEGER;

for some constant **MAXSIZE**.

Specifying the parameters to **SORT** and determining the types involved are part of designing the *interface* or *calling sequence* for the **SORT** procedure. Completing this specification early in the design phase is very important in the computing industry, particularly if it is part of a large programming project. One of the important advantages that comes from making such design decisions early is that it is then possible for the implementation of the **SORT** procedure to proceed in parallel with the application programs that will use it.

7-1   Selection Sorting

With the design of the **SORT** procedure interface out of the way, we can return to the algorithmic problem of sorting the array

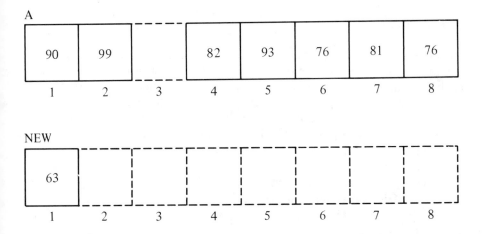

In selection sorting, the first step is finding the smallest value (in this case, 63), since this is certain to be the first value in the sorted array. If we copy this value to a separate array called **NEW** and conceptually "remove" it from the original, we are left with the following picture:

Repeating this process until each element has been moved into **NEW** results in a fully sorted array. The fact that each step in the sorting procedure consists of "selecting" the next element gives rise to the term "selection sort." For many people, this is a familiar operation and is used, for example, in sorting a hand of cards. After the deal, the cards are held in one hand and then selected in order by the other, resulting in a properly sequenced hand.

As described, this algorithm is not particularly well suited to implementation, primarily because the idea of "removing" an entry from an array is not one that is easily expressed in most programming languages. For this reason, it is helpful to modify the algorithm so that, instead of removing an element, we simply exchange it with the value in the desired position. Thus, in the original array, we must first find the smallest value (indicated by the "S" in the diagram below):

90	99	63	82	93	76	81	76
1	2	3	4	5	6	7	8
^		^					
I		S					

and then exchange it with the value in the first position.

63	99	90	82	93	76	81	76
1	2	3	4	5	6	7	8
^		^					
I		S					

At this point, the first element is correctly in place and the remaining items yet to be sorted are in positions 2 through 8. If we repeat this operation for each remaining position, we will end up with all the elements in order.

In point of fact, we will not actually need to perform all eight cycles here. If positions 1 through 7 contain the correct values, then the one remaining value must belong in position 8. Thus, our more general algorithm need only exchange elements into the positions from 1 up to N–1.

If we choose a reasonable way of breaking up the problem into appropriate subsidiary procedures, this algorithm is readily translated into Pascal. Developing the **SORT** procedure using the principles of top-down design makes the problem easier still. The essence of the algorithm is represented by the following Pascal program:

```
PROCEDURE SORT(VAR A : INTARRAY; N : INTEGER);
    VAR
    I, S : INTEGER;
    BEGIN
    FOR I := 1 TO N–1 DO
      BEGIN
      S := FINDSMALLEST(A, I, N);
      SWAP(A[I], A[S])
      END
    END;
```

Translated into English, this can be interpreted as*

For each position I between 1 and N–1:

1. Call the function **FINDSMALLEST** to determine the position of the smallest element in A in the index range I to N. Assign that position number to S.
2. Swap the values in A[I] and A[S].

The remaining procedures are equally easy to code and should be familiar from any introductory programming course.

```
PROCEDURE SWAP(VAR X, Y : INTEGER);
  VAR
   TEMP : INTEGER;
  BEGIN
   TEMP := X;
   X := Y;
   Y := TEMP
  END;


FUNCTION FINDSMALLEST(VAR A : INTARRAY;
                      P1,P2 : INTEGER) : INTEGER;
  VAR
   S, I : INTEGER;
  BEGIN
   S := P1;
   FOR I := P1+1 TO P2 DO
    IF A[I] < A[S] THEN S := I;
   FINDSMALLEST := S
  END;
```

In order to have any idea whether or not this is a reasonable sorting algorithm, we must determine a metric for judging the quality of its operation. Although we may end up being interested in other characteristics of the program operation (such as how much space it uses or how well it operates on arrays that are already partially ordered), computational complexity is certainly of considerable importance.

To calculate the complexity order, the most common approach is to choose some operation in the program and count the number of times it is performed.

---

*In this chapter, the selection sort algorithm is coded iteratively using a **FOR** loop to cycle through each position. Alternatively (see Exercise 7-2), this could also be viewed as a recursive procedure which correctly positions the first item and then sorts the rest of the list recursively.

In order to ensure that this measure is at all useful, it is important to select an operation that is executed at least as often as any other. Thus, the usual approach is to select some operation from the innermost loop. Here, since the body of the loop within **FINDSMALLEST** is executed several times during each cycle of the loop in **SORT,** it is conventional to choose some operation there, such as the comparison in the statement

$$\textbf{IF A[I]} < \textbf{A[S] THEN S := I;}$$

Using the eight-element example presented earlier, we can compute the number of comparisons required for each cycle of the **SORT** program. On the first cycle, **FINDSMALLEST** is called with **P1** equal to 1 and **P2** equal to 8, which means that the comparison will be performed seven times—once for each cycle of the loop

$$\textbf{FOR I := P1+1 TO P2 DO . . .}$$

On the second cycle of **SORT,** there will be six comparisons, five on the third, and so on. Thus, the total number of comparisons in the eight-element example is

$$7 + 6 + 5 + 4 + 3 + 2 + 1 = 28$$

More generally, if the original array contains N elements, the number of comparisons is given by

$$1 + 2 + 3 + \cdots + (N-1)$$

We can simplify this formula by taking advantage of the result from Chapter 2, which establishes that

$$1 + 2 + 3 + \cdots + N = \frac{N(N+1)}{2}$$

Thus, the number of comparison operations required to sort an array of size N is

$$\frac{(N-1)N}{2}$$

or, in polynomial form

$$\tfrac{1}{2}N^2 - \tfrac{1}{2}N$$

Since the second term in this formula becomes relatively insignificant for large values of N, the complexity of selection sort is simply

$$O(N^2)$$

The derivation of this result is almost exactly the same as that of the nested loop example presented in Chapter 2. In particular, the computational characteristics are just as bad. The dimensions of the problem are indicated by the following table:

N	Number of comparisons
10	45
20	190
50	1,225
100	4,950
1000	499,500
10000	49,995,000

From this table, we see that the number of comparisons required grows much more quickly than the number of items to be sorted. For small values of N, this is not particularly significant, but, as N grows, this soon becomes a matter of no small concern. For example, if the computer we are using can perform 100,000 comparisons per second, sorting 10,000 data items using this method will take nearly 8 1/2 minutes for the comparison operations alone. Even this may not seem completely preposterous (to anyone who is not paying directly for computer time), but it quickly gets worse. For 100,000 items (which is entirely within the range of a company working with a large bulk mailing list), the program will require nearly 5,000,000,000 comparisons. This would require almost *fourteen hours,* given the same rate of computation.

As these figures indicate, this algorithmic inefficiency makes selection sorting unacceptable for most applications involving large collections of data. We can do better than this by adopting a recursive approach.

7-2 Merge Sorting

In looking for a recursive solution, we must first discover what characteristics of sorting make it a recursive problem. To do this, it helps to review the conditions that recursive problems must satisfy:

1. There must be some way to break large problems down into simpler instances of the same problem.

2. Assuming that each of those subproblems can be solved by successive applications of the recursive procedure, there must be some way to generate a solution to the original problem from the solution to each of these smaller parts.
3. We must be able to determine a set of simple cases which can be solved directly, without any further decomposition.

To be sure that we are heading in a profitable direction, it often pays to consider the simple cases first. Here, given that the problem is sorting an array of size N, the easiest case is when N = 1. When this occurs, we have a list of a single element, which, considered all by itself, is already sorted. Thus, for recursive sorting, the simple case requires no work at all.

Once this is out of the way, we can return to the more complicated task of subdividing the problem. Given a large array, we can break it up into smaller arrays in several different ways. When we are trying to reduce the computational complexity of an algorithm, it usually helps to divide the problem into components which are approximately equal in size. In general, this approach takes maximum advantage of the divide-and-conquer strategy and ensures that the size of the resulting subproblems diminishes quickly as the recursive solution proceeds. Thus, in this case, the most promising approach is to (1) divide the array in half, (2) sort each of the subarrays using a recursive call, and (3) find some way to reassemble the two subarrays into a fully sorted array.

Before tackling the details of this algorithm, however, we must recognize that the use of this recursive approach requires some changes in the structure of the **SORT** procedure itself. Before we settled on the algorithm to be used in the implementation, we designed the **SORT** procedure to take *two* arguments as indicated by its header line

PROCEDURE SORT(VAR A : INTARRAY; N : INTEGER);

In the case of the selection sort algorithm from the previous section, this was quite sufficient, because the algorithm was designed to work with the entire array all at once.

In the recursive implementation, however, the algorithm is predicated on an ability to divide the complete array into subarrays, which are handled individually. For example, in sorting an array with eight elements, we will first call **SORT** recursively on elements 1 through 4, and then call it again for elements 5 through 8. Thus, the **SORT** procedure must be given *both* endpoints of the desired range rather than the number of elements alone.

Having made this discovery, we are tempted to change the procedure header for **SORT** to

PROCEDURE SORT(VAR A : INTARRAY; LOW, HIGH : INTEGER);

Unfortunately, this approach is extremely poor from the point of view of project management. Presumably, other programmers have written application programs which assume that the **SORT** procedure will conform to its design specifications. Altering the calling sequence at this point requires all those programs to change.

A better solution to this problem is to leave the calling sequence for **SORT** alone and introduce a new procedure which adds in the extra argument. Thus, the **SORT** routine itself is simply

```
PROCEDURE SORT(VAR A : INTARRAY; N : INTEGER);
BEGIN
    MERGESORT(A, 1, N)
END;
```

All of the actual work for the recursive solution is performed by the new **MERGESORT** procedure, which is designed to handle the more general case of arbitrary subarrays. This technique of using two procedures, one which provides a simple external interface and a second (typically with more parameters) which implements the actual algorithm, is quite important in recursive programming and will be used again in subsequent chapters.

Even though we have yet to specify the details of the complete algorithm, we are in a position to outline the code for the **MERGESORT** procedure, in accordance with the structure suggested by the informal algorithmic description:

```
PROCEDURE MERGESORT(VAR A : INTARRAY;
                        LOW, HIGH : INTEGER);
BEGIN
    IF there is any work to do THEN
        BEGIN
            Find the middle element and mark that as MID;
            Recursively sort the subarray from LOW to MID;
            Recursively sort the subarray from MID + 1 to HIGH;
            Reassemble the two sorted subarrays END
        END
END;
```

To handle the simple case correctly, we must check if there is more than a single element in the **LOW** to **HIGH** range. This is true only if **LOW** is less than **HIGH,** so that the necessary **IF** test is simply

IF LOW < HIGH THEN

The remainder of the **MERGESORT** procedure is executed only if there is more than a single element, since a subarray with only one element requires no sorting operations at all.

If there is more than a single element, we must determine the middle of the range by computing the average of **LOW** and **HIGH.** Once this midpoint is calculated, the resulting subarrays can be sorted by making recursive calls to **MERGESORT,** updating the arguments appropriately.

The final step in the algorithm consists of reassembling these two subarrays so that **A** contains all the elements in their proper order. In computer science, the operation of combining two or more sorted arrays into a single larger array is called *merging*. Since merging involves quite a number of steps, it is useful to separate this phase of the algorithm by defining a separate **MERGE** procedure. This improves the overall program structure and makes it possible to complete the definition of the **MERGESORT** procedure itself:

```
PROCEDURE MERGESORT(VAR A : INTARRAY;
                         LOW, HIGH : INTEGER);
   VAR
     MID : INTEGER;
   BEGIN
     IF LOW < HIGH THEN
       BEGIN
         MID := (LOW + HIGH) DIV 2;
         MERGESORT(A, LOW, MID);
         MERGESORT(A, MID+1, HIGH);
         MERGE(A, LOW, MID, HIGH)
       END
   END;
```

Before we turn to the details of the **MERGE** procedure, it is useful to trace the **MERGESORT** operation on a specific example. If we run **MERGESORT** on the array **A** introduced earlier in the chapter, the first step consists of dividing **A** into two halves by regarding the elements in positions 1 through 4 and 5 through 8 as conceptually distinct arrays.

A

90	99	63	82	93	76	81	76
1	2	3	4	5	6	7	8

Drawing courage from the recursive leap of faith, we assume that these simpler problems can be solved by making recursive calls to the **MERGESORT** procedure. This assumption means that we can reach the configuration

A

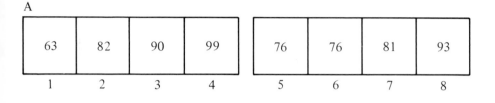

This brings us to the point at which we must apply the **MERGE** operation. Merging two sorted arrays into a single larger array turns out to be considerably simpler than sorting in terms of its computational efficiency. This increase in efficiency comes from the fact that, in order to choose the next element for the combined array, it is necessary to look at only one element from each of the subarrays.

To keep track of our progress in each of the subarrays, we will introduce the variables **I1** and **I2** and initialize them so that they indicate the beginning of each subarray:

A

The general procedure for performing the merge operation consists of (1) choosing the smaller of the elements at positions **I1** and **I2**, (2) copying that element into the next position in a temporary array named **TEMP,** and (3) incrementing either **I1** or **I2**, depending on which element was selected.

On the first cycle, 63 is less than 76, so the first data value is chosen from position **I1**. This value is then copied into the first element in **TEMP**, and **I1** is incremented to reference the next element in the first subarray. This gives rise to the following situation:

A

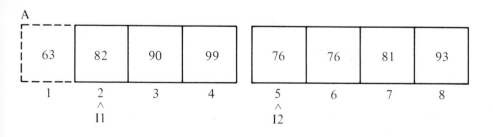

TEMP

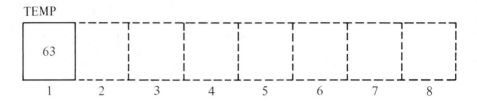

For each of the next three cycles, the value to be chosen comes from the second subarray, so that the **I2** position index will move up to 8 as the elements in positions 5 through 7 are copied into **TEMP:**

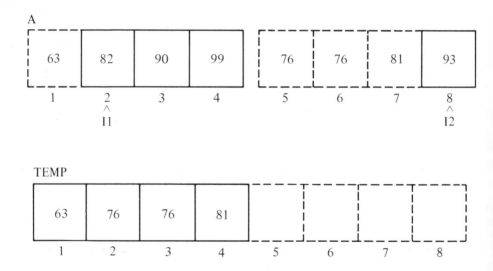

When the remaining elements have been processed in the same way, **TEMP** will contain the completely sorted array.

TEMP

63	76	76	81	82	90	93	99
1	2	3	4	5	6	7	8

Conceptually, the **MERGE** algorithm therefore has the following form:

```
PROCEDURE MERGE(VAR A : INTARRAY;
                         LOW,MID,HIGH : INTEGER);
VAR
  TEMP : INTARRAY;
  I, I1, I2 : INTEGER;
BEGIN
  I1 := LOW;
  I2 := MID + 1;
  FOR I := LOW TO HIGH DO
    BEGIN
      IF (A[I1] should be chosen before A[I2])
        THEN BEGIN TEMP[I] := A[I1]; I1 := I1 + 1 END
        ELSE BEGIN TEMP[I] := A[I2]; I2 := I2 + 1 END
    END;
  FOR I := LOW TO HIGH DO A[I] := TEMP[I]
END;
```

In this program, I1 and I2 are set to the position numbers corresponding to the beginning of each subarray. From there, we go through each of the positions in TEMP and fill it with either the value A[I1] or A[I2] as appropriate. Whichever one we choose, we must also increment that index value so that it refers to the next element in the subarray.

Of course, the implementation is not complete until we replace the English phrase

$$A[I1] \text{ should be chosen before } A[I2]$$

with Pascal code to accomplish the same job. In the usual case, the condition that belongs in this position is simply

$$A[I1] < A[I2]$$

This, however, does not take into account the fact that we might have exhausted the elements in one of the two parts. A more correct interpretation for the intent of the above English phrase is indicated by the following case-by-case analysis:

> If we are past the end of PART1, choose A[I2].
> If we are past the end of PART2, choose A[I1].
> Otherwise, choose the smaller of A[I1] and A[I2].

One of the easiest ways to represent this in Pascal is to introduce a Boolean variable CHOOSE1 which is TRUE exactly when the element A[I1] should be chosen. We set **CHOOSE1** using Pascal code which represents the set of tests above:

```
IF I1 > MID
  THEN CHOOSE1 := FALSE
  ELSE IF I2 > HIGH
    THEN CHOOSE1 := TRUE
    ELSE CHOOSE1 := (A[I1] < A[I2]);
```

Note that the last **ELSE** clause sets **CHOOSE1** to **TRUE** or **FALSE** depending on whether or not **A[I1]** is smaller than **A[I2]**. Thus, the complete implementation of **MERGE** is

```
PROCEDURE MERGE(VAR A : INTARRAY;
                LOW, MID, HIGH : INTEGER);
VAR
  TEMP : INTARRAY;
  I, I1, I2 : INTEGER;
  CHOOSE1 : BOOLEAN;
BEGIN
  I1 := LOW;
  I2 := MID + 1;
  FOR I := LOW TO HIGH DO
    BEGIN
      IF I1 > MID
        THEN CHOOSE1 := FALSE
        ELSE IF I2 > HIGH
          THEN CHOOSE1 := TRUE
          ELSE CHOOSE1 := (A[I1] < A[I2]);
      IF CHOOSE1
        THEN BEGIN TEMP[I] := A[I1]; I1 := I1 + 1 END
        ELSE BEGIN TEMP[I] := A[I2]; I2 := I2 + 1 END
    END;
  FOR I := LOW TO HIGH DO A[I] := TEMP[I]
END;
```

At this point, we have a working **MERGESORT** procedure. As of yet, however, we have not demonstrated that it provides any increase in performance over selection sorting. To verify this claim, we must analyze the computational complexity of the merge sort algorithm.

Pictorially, we can get some idea of the computational requirements by considering a diagram of the complete recursive decomposition:

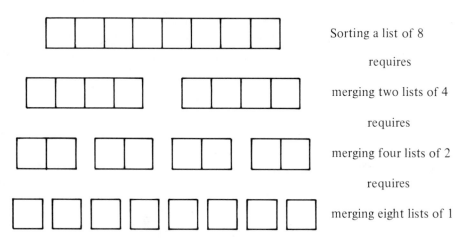

Sorting a list of 8

requires

merging two lists of 4

requires

merging four lists of 2

requires

merging eight lists of 1

At each level, all the real work is done by the **MERGE** routine. Whenever we call **MERGE** with two sublists whose combined size is K, the main internal loop runs for exactly K cycles. This suggests that the operation of **MERGE** is linear in the size of the sublist. We may be a little concerned by the fact that there are two independent loops requiring K cycles and that many operations are required to carry out the first loop, but this can only increase the running time of **MERGE** by a constant factor. The behavior of **MERGE** remains linear.

We can further generalize the above diagram by allowing the size of the original list to be some arbitrary value N. At each level, we cut the size of the subarrays in half. Since we carry this out until the size of each subarray is 1, there must be log N levels of recursive call below the original problem level, as shown in the following diagram:

Sorting a list of size N
requires
merging two lists of size N/2
requires
merging four lists of size N/4
requires
. . . .
requires
merging N lists of size 1

log N
times

The key insight revealed by this diagram is that the same amount of work is required at each level of the recursive decomposition. Each successive level requires merging twice as many sublists, each of which is half the previous size. Since the **MERGE** operation is linear in the number of elements to be merged, dividing the size of the list in half should also cut the time requirements in half. However, since we have twice as many **MERGE** operations to perform, the work required at each level remains constant. Thus, the total work is proportional to the cost of the **MERGE** operation for the first level multiplied by the number of levels. This indicates that the complexity of merge sorting must be

$$O(N \log N)$$

Just looking at the complexity formula, however, is not altogether convincing. To show that this is an enormous improvement, it helps to go back to the earlier table which indicated the performance for selection sort given various values of N:

N	N^2	N log N
10	100	33
20	400	86
50	2,500	282
100	10,000	664
1000	1,000,000	9,966
10000	100,000,000	132,877

In reality, we have cheated a bit here. Reducing the complexity calculation to big-O notation invariably means that we have ignored much of the underlying detail. In particular, big-O notation throws away the coefficient on the leading term. From our complete analysis, we remember that selection sorting required approximately

$$\tfrac{1}{2}N^2$$

steps. Thus, the figures in the center column are too high by a factor of two. Similarly, the figures in the last column are probably much too low. The **MERGE** procedure requires two complete cycles through the data, once to enter values into the **TEMP** array and once to copy it back. Furthermore, the main loop inside **MERGE** is considerably more complicated than many of the operations required for selection sorting. Measured in the same time units as selection sort, it may be that merge sorting requires as much as

$$5N \log N$$

although this is clearly just a guess. To get exact values, it would be necessary to count instructions or measure the required time. Using this guess as a more representative indication of the actual time requirements for merge sort gives the following table:

N	$N^2/2$	5N log N
10	50	166
20	200	432
50	1,250	1,411
100	5,000	3,322
1000	500,000	49,829
10000	50,000,000	646,385

This table shows a more interesting situation. As long as N is reasonably small, merge sorting takes more time than selection sorting. As soon as N gets larger than about 60, however, merge sorting begins to pull into the lead. The bottom line remains impressive. To sort 10,000 items using merge sorting requires only 1.3% of time that selection sorting would require—a savings of 98.7%. Compared with the fourteen hours required to sort 100,000 items using selection sort, merge sort should be able to do the job in just under a minute and a half.

As an aside, there is an alternative method of calculating the complexity order for merge sort that does not require looking at the complete recursive decomposition and is therefore more consistent with the holistic view of recursion. Except when the easy cases are encountered, the **MERGESORT** routine performs the following operations:

> **MID := (LOW + HIGH) DIV 2;**
> **MERGESORT(A, LOW, MID);**
> **MERGESORT(A, MID+1, HIGH);**
> **MERGE(A, LOW, MID, HIGH)**

If we use N to represent the number of elements in the range **LOW** to **HIGH**, we can immediately conclude that

> <run time needed to sort N items>
>   = 2 × <run time needed to sort N/2 items>
>   + <run time needed to merge N items>

If we use sorttime(N) as as abbreviation for "run time needed to sort N items" and remember that the time required to merge N items is proportional to N, we can rewrite this as

$$\text{sorttime}(N) = 2 \times \text{sorttime}(N/2) + N$$

This is a recursive definition and requires a simple case, which turns out to be the same as that used in the **MERGESORT** procedure itself.

$$\text{sorttime}(1) = 0$$

This means that we could even write a Pascal function to compute the running time as

```
FUNCTION SORTTIME(N : INTEGER) : INTEGER
  BEGIN
    IF N<= 1
      THEN SORTTIME := 0
      ELSE SORTTIME := 2 * SORTTIME(N DIV 2) + N
  END;
```

Although it is not immediately obvious, this recursive formulation (at least for powers of 2) is equal to N log N. This fact can be demonstrated using mathematical induction and is discussed further in Exercise 7-3.

Bibliographic Notes

As noted earlier in the chapter, sorting algorithms are of enormous practical importance in computer science. For this reason, considerable research has been dedicated to the problem of sorting techniques, and the list of sources in this area is extremely large. The landmark text is Volume III of Don Knuth's *Art of Computer Programming* (*Sorting and Searching*) [1973]. This text also includes extensive bibliographic references associated with each of the algorithms presented. Sedgewick [1983] also contains an excellent discussion of the sorting techniques presented here.

Exercises

7-1. Deal a hand of thirteen playing cards and sort it using the merge sort procedure described in the text. Use this experience to defend the notion that, even though merge sorting may be computationally superior, selection sorting still has its place.

7-2. Recode the "selection sort" algorithm from section 7-1 so that it operates recursively.

7-3. Assuming that N is always a power of two, prove that the recursive formula

$$\text{sorttime}(N) = \begin{cases} 0, & \text{if } N = 1 \\ 2 * \text{sorttime}(N/2) + N & \text{otherwise} \end{cases}$$

can be simplified to

$$\text{sorttime}(N) = N \log N$$

The proof of this proposition can be simplified considerably by using a more powerful form of mathematical induction than the technique of *simple induction* presented in Chapter 2. The principle of *strong induction* is based on the observation that, as we build upward from the simple case, we generate an entire sequence of theorems. For example, in trying to prove that a formula is true for $N = 5$, we should be able to draw upon the fact that it is true for $N = 1, 2, 3$, and 4, and not just the $N = 4$ assumption allowed by the simple induction.

A proof using strong induction proceeds as follows:

1. Prove that the base case holds by establishing the correctness of the formula when $N = 1$.
2. Assume that the formula is true *for all values* from 1 up to some arbitrary number N.
3. Using that hypothesis, establish that the formula holds for $N + 1$.

7-4. Even though merge sorting is computationally more efficient than many of its competitors, it has its problems. First, since the **MERGE** procedure uses a temporary array, the algorithm will require twice as much storage as an algorithm which sorts the array "in place." Moreover, merge sort does have a relatively high constant of proportionality. This is reflected in the observation that, in practice, merge sorting performs relatively poorly on small lists.

To get around each of these problems, other sorting algorithms have been developed. One of the most useful is the "Quicksort" algorithm, developed by C. A. R. Hoare in the early 1960s. Conceptually, the Quicksort algorithm operates as follows:

1. Arbitrarily select some element in the array and call it the *pivot* value. For example, we could always choose the first element for this purpose.
2. Go through the entire array, rearranging the elements so that the array is divided into two subarrays separated by the pivot value. All elements which come before the pivot must be smaller than the pivot, while those which follow must be at least as large. This operation is usually referred to as *partitioning* the array.
3. Recursively sort the subarrays that precede and follow the pivot, respectively.

This process can be illustrated in the context of the array used throughout the examples from this chapter. Initially, the array contains the elements

90	99	63	82	93	76	81	76
1	2	3	4	5	6	7	8

with the first element chosen to be the pivot. After partitioning the array, we hope to reach a configuration like

76	63	82	76	81	90	93	99
1	2	3	4	5	6	7	8

$$\underset{\underset{P}{\wedge}}{}$$

where the pivot has moved into the interior of the array. From here, sorting the subarrays before and after the pivot results in a completely sorted list.

It is important to note that the actual array we get after this step depends entirely on how the partition operation is implemented. The only requirement for partition is that the elements before and after the pivot value in the partitioned list are correctly oriented with respect to the pivot. There are no requirements as to how they get there or how they are oriented with respect to each other.

Given this flexibility, there are many acceptable methods for accomplishing the partition phase of Quicksort. One of the simplest is attributed to Nico Lomuto and operates as follows:

1. Let the variable **BOUNDARY** hold the position which separates the values smaller than the pivot from those which are not. Originally, **BOUNDARY** is set to the first position in the array where the pivot element resides.
2. Look at each remaining item in the array. If this element is smaller than the pivot value, we would like to move it toward the beginning. This is accomplished by (a) switching that element with the first element *after* the current **BOUNDARY** and (b) moving **BOUNDARY** over by one.
3. Switch the first element (i.e., the pivot) with the value at position **BOUNDARY**. Return this position as the value of the pivot index, denoted by **P** in the diagrams above.

Using the algorithmic description above, write a Pascal implementation of Quicksort. If you need further explanation of the procedure, consult Jon Bentley's "Programming Pearls" section of the April 1984 issue of *Communications of the ACM*.

Intelligent
Algorithms

A mighty maze! but not without a plan
>   —*Alexander Pope,* An Essay on Man, *1733*

Of those areas that make up computer science, there is one which has captured more popular attention in recent years than any of the rest—*artificial intelligence*. The idea that a machine could be endowed with the capacity for thought is enticing to some and profoundly disturbing to others.

Prior to the development of modern computers, the debate over artificial intelligence and its potential was principally theoretical in character, giving the entire question a certain philosophical quality. Since that time, however, technology has improved to the point where computers can perform many activities which seem to require a considerable level of "understanding." For example, the MACSYMA system (developed by Joel Moses and others at M.I.T.) can apply complex mathematical techniques in ways that seem to require insight or cleverness. The MYCIN system (by Edward Shortliffe at Stanford) integrates information about a patient's medical history, current symptoms, and test results to produce a diagnosis and plan for treatment.

Even at the level of the personal computer, many programs have been written which, to some extent, "understand" English sentences. Of these, the best known examples are found in "adventure" games, where one can order the computer, for example, to

> Throw the treasure to the troll, then cross the bridge.

While the context may seem too limited and esoteric to be useful, the algorithms used in these games are very much the result of artificial intelligence research.

Rather than attempting to survey artificial intelligence as a field, this chapter illustrates the use of recursion as a decision-making mechanism in two problem domains: (1) finding a path through a maze and (2) choosing a move in a strategy game.

8-1 Backtracking through a Maze

As at most conferences, the participants at the 1979 National Computer Conference in New York City spent much of their time attending technical sessions or surveying the new tools of the trade. In that year, however, there was a special event that made the conference much more exciting, for this was the setting for the first "Amazing Micro-Mouse Maze Contest," sponsored by IEEE/Spectrum. The engineering teams that entered the contest had built mechanical "mice" in accordance with a detailed set of rules published in 1977. The mission of these cybernetic mice was to perform the classic demonstration of rodent intelligence—running a maze.

Although the idea of building a mechanical mouse brings up a wide range of interesting engineering problems (such as how it should move or notice walls), the programming aspect of the problem is interesting in its own right. Suppose, for example, that we want to write a Pascal program to solve mazes like the one pictured below:

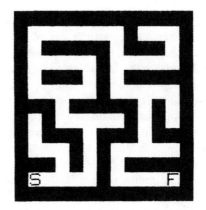

Such a maze consists of a rectangular array of smaller squares, which can be either empty (representing a corridor section) or filled in (representing a wall). In addition, two of the squares are labeled with the letters 'S' and 'F', representing the start and finish squares, respectively.

From the programming point of view, the problem here is to find some way to generate a solution path. More specifically, we can try to construct a program that, given a representation of the above maze as input, will generate a diagram in which one of the possible solution paths is marked.

In order to develop an algorithm for solving a maze, it will be necessary to adopt a systematic approach that explores every possible path until a solution is found. Since some of the paths will lead to dead ends, the process of exploring those paths will necessarily require "backing up" when no further progress can be made. For example, if we begin by going two squares forward (marking our path as we go), we reach a point where we must choose between two possible paths. Such a square will be called a "choice point" in the subsequent discussion.

From here, we must choose one path for an initial exploration, and, if that proves unsuccessful, return to this choice point to try the other. Looking down on the maze from above, it is clear that we should continue east and not north, since the latter choice runs quickly into a dead end. Any program that we write to solve this problem, however, will be forced to attempt this path until that dead end is actually discovered.

If the program is unlucky enough to choose the north path first, it will explore ahead until it reaches the end of that path and notices that it can no longer proceed.

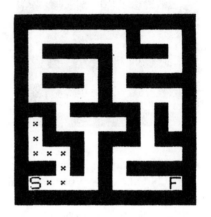

When this occurs, the program must enter a backtracking phase, during which it moves backward along its original path until it reaches an earlier choice point. Here, for example, it will back up to the first junction where it made its original decision to go north. From here, it will try the east path from this square and proceed on toward an eventual solution. Algorithms that use this approach of testing possible solutions and returning to previous decision points when such attempts fail are generally referred to as *backtracking algorithms*.

The general idea outlined above can be turned into an informal algorithmic description as follows:

1. Start initially at the square labeled by 'S'.
2. Drawing an 'X' on each square that you cross, follow the current path until you encounter any of the following:
   a. If you hit the 'F' marker, you are finished and the correct path is indicated by the chain of 'X' markers currently in place.
   b. If you hit a choice point, consider each of the possible directions in some fixed order. Repeat this process along that path until you are told to return to this square. When this occurs, choose the next possible direction. If all the possible directions have been attempted, this entire area represents a dead end, and you should proceed to case (d).
   c. If you hit a square which has already been marked, then the maze contains a loop. To avoid going around that loop forever, it is essential to consider this case as one of the "failure" cases and follow the instructions for case (d).
   d. If you hit a dead end, return to the previous choice point and try the

next possible direction. As you return, erase the 'X' markers along the path. If there are no previous choice points, the maze has no solution.

Although this algorithmic description is readily understood, it is based on several notions which are difficult to integrate into a working program. For example, the phrase "follow the current path" seems straightforward enough, but turns out to be somewhat more difficult to implement than it sounds. Similarly, the idea that we can "return to the previous choice point and try the next possible direction" is more easily expressed in English than in Pascal.

We can simplify the process of translating this algorithm into Pascal by making it a little *less* clever. As described above, the algorithm must be able to tell if a given square in the maze is a dead end, a choice point, or a "normal" square along some path. First of all, we can eliminate the need for dead-end detection simply by letting the program "run into the wall." If the program takes a step forward and finds that it has reached a wall square, then it was not possible to proceed in that direction. We can further reduce the complexity by considering *every* square in the maze to be a choice point with exactly four possible directions. Some of these, of course, run directly into walls, but this situation can be handled as part of the standard course of events.

It is easiest to illustrate these changes in the context of a very simple maze. Except for its size, this maze is similar to the earlier example. The only structural difference is that the start position has been replaced by a small diamond which indicates the position of the "current square."

As noted above, we will consider each square in the maze to be a choice point with four possible exit paths leading north, east, south, and west. If we select these in that order, the next position we will consider is the one we reach by moving one square to the north. This puts us in the following state:

From here, the process starts all over again, and the program will once again try each of the four directions starting with north. In this case, the north path is less successful and leads to the following configuration:

In this figure, the fact that the diamond indicating the current square is now sitting on top of a wall means that we are unable to proceed in this direction and must retreat. We return to the position

and try moving east. Unfortunately, this is no better, so we again retreat and try south. Here we encounter a square which is already marked, indicating that we have doubled back onto our previous path. Since there is no point in reexploring old territory (particularly since we are likely to get stuck in a loop this way), we reject this direction as well and try the remaining one. When moving west proves to be equally unsuccessful, we know that we have reached a dead end. This forces us to back up one more square, erasing the mark from this one.

When we do this, we are back in the configuration

ready to try moving east. On the whole, this proves more profitable, and, after banging into many more walls, we reach the final configuration:

To implement this algorithm in Pascal, we must find a way to manage the backtracking operation. Somehow, we must keep track of where we have been in the maze and what directions we have tried at each choice point. This is where recursion comes in handy.

In order to see how recursion can be applied to this problem, we must frame this solution technique so that solving a maze can be represented in terms of solving simpler ones. To do this, it is again helpful to consider a specific maze.

Throughout the earlier discussion of this problem, whenever we have looked at a potential solution, we have followed that possibility all the way to a conclusion. Thus, if we were to try moving north in this maze, we would keep on pursuing that path until we discovered the dead end around the corner. Thinking in this way, however, is not ordinarily conducive to finding the recursive structure of the problem. Here we must simplify the problem one step at a time.

The insight necessary to design the recursive algorithm comes from the following observation:

Starting from the position

the entire maze has a solution if, and only if, at least one of the following mazes can be solved:

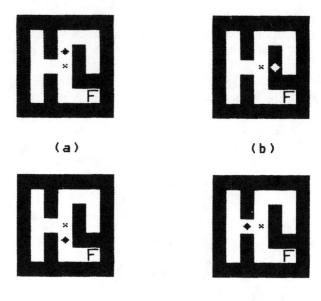

(a)   (b)

(c)   (d)

Think about this assertion carefully. In making these new mazes, we have transformed the original problem into four new ones by replacing the current position with an 'X' marker and advancing in each of the four compass directions. Using recursion, we will try to solve each subproblem in turn. If any of these yields a solution, then we have a solution from our initial configuration. If none do, then the maze is unsolvable from the given initial state. Moreover, if we mark each blank square we encounter with an 'X' and then unmark it only after we have tried all four directions, then the final maze will show the solution path.

Each of these new subproblems is simpler than the original because the number of open corridor squares has been reduced by one, since the square we previously occupied has been replaced by an 'X' marker. For this recursion to be effective, this simplification must eventually result in simple cases that allow the recursion to terminate. For the most part, these are the same as those given in the informal algorithmic description.

1. If the current square is marked with an 'F', then we have found a solution path.
2. If the current square is a wall square or is already marked with an 'X', then no path exists in this direction.

Before we turn to coding this in Pascal, we need to choose an appropriate data structure. For the maze itself, a two-dimensional matrix of characters seems appropriate:

```
VAR
    MAZE : ARRAY [1..MAXWIDTH, 1..MAXHEIGHT] OF CHAR;
```

If **X** and **Y** represent our current coordinates within the maze, then **MAZE [X,Y]** contains a character which indicates the type of square. Those characters are selected according to the following code:

Space	An open passageway
'#'	A wall square
'F'	The finish square
'X'	A marked square

The start square will not be explicitly represented in the matrix, since it is used only to establish the current position at the beginning of the algorithm.

To put the entire program in perspective, we will assume that we have a procedure **READMAZE,** which (1) reads in the **MAZE** matrix and (2) sets the variables **STARTX** and **STARTY** to indicate the coordinates of the initial square. Once this data structure is initialized, we must then call a subprogram which applies the recursive strategy above to solve that maze. Since that subprogram must return an indication of whether or not it was able to find a solution path, it makes sense to design it as a function which returns **TRUE** or **FALSE** to indicate success or failure. On a successful return from **FINDPATH,** the program can simply print the current **MAZE** structure since that will show the correct path as a line of 'X' markers. Otherwise, the program should indicate that no solution exists. Thus, the main program has the following general structure.

```
BEGIN
    READMAZE;
    IF FINDPATH(STARTX, STARTY)
        THEN PRINTMAZE
        ELSE WRITELN('There is no solution.')
END.
```

We are now in a position to sketch an outline of the **FINDPATH** routine, as shown on the top of page 112. Turning the internal English description into legal Pascal is a bit tricky, given the restrictiveness of the control structures provided by Pascal. Structurally, we need a loop that exits when either (1) a solution is found or (2) all four directions have been tried. In Pascal, this requires a **WHILE** loop that updates the direction variable explicitly as part of the body.

```
FUNCTION FINDPATH(X, Y : INTEGER) : BOOLEAN;
BEGIN
  CASE MAZE(X,Y) OF
    'F'      : FINDPATH = TRUE;
    '#', 'X' : FINDPATH := FALSE;
    ' '      :
      BEGIN
        Set MAZE[X,Y] to 'X';
        For each possible direction, call
          FINDPATH recursively until a
          solution is found;
        If none is found then set
          MAZE[X,Y] back to ' '
      END
  END
END;
```

```
FUNCTION FINDPATH(X, Y : INTEGER) : BOOLEAN;
VAR
  DIR : DIRECTION;
  FOUND : BOOLEAN;
BEGIN
  CASE MAZE(X,Y) OF
    'F'      : FINDPATH := TRUE;
    '#', 'X' : FINDPATH := FALSE;
    ' '      :
      BEGIN
        MAZE[X,Y] := 'X';
        FOUND := FALSE;
        DIR := NORTH;
        WHILE (NOT FOUND) AND (DIR <> FAILED) DO
          BEGIN
            CASE DIR OF
              NORTH : FOUND := FINDPATH(X, Y-1);
              EAST : FOUND := FINDPATH(X+1, Y);
              SOUTH : FOUND := FINDPATH(X, Y+1);
              WEST : FOUND := FINDPATH(X-1, Y)
            END;
            DIR := SUCC(DIR)
          END;
        IF NOT FOUND THEN MAZE[X,Y] := ' ';
        FINDPATH := FOUND
      END
  END
END;
```

Although we could use, for example, the integers 1 through 4 to indicate the four compass points, this application is ideal for the introduction of a user-defined scalar type which will serve as the type for the loop control variable:

DIRECTION = (NORTH, EAST, SOUTH, WEST, FAILED);

The value **FAILED** is included in the list so that the **WHILE** loop can test if the list of directions is exhausted. In that loop, we will start by setting the variable **DIR** to **NORTH**. After each cycle, we will use the built-in function **SUCC** to increment **DIR** so that it specifies the next direction:

DIR := SUCC(DIR)

This makes it possible to complete the Pascal implementation of **FIND-PATH**, as shown on the bottom of page 112.

8-2   Lookahead Strategies

In 1769, the Hungarian inventor Wolfgang von Kempelen developed a machine that, from all outward appearances, represented a remarkable advance in engineering. Christened "The Turk," the machine consisted of a life-sized figure (complete with turban and Turkish attire) seated behind a chessboard. Accompanied by the requisite grinding of gears and sliding of rods, The Turk would play exhibition chess matches during which it displayed great proficiency at the game. Brought to the United States in 1826 by Johann Maelzel (better known for his invention of the metronome), The Turk was given to the Chinese Museum in Philadelphia, where it was eventually destroyed by a fire which devastated much of the museum collection.

On closer examination, The Turk's success as a chess player was revealed as somewhat less than remarkable, since it depended rather intimately on the human chess player hidden inside. One of the most detailed accounts of this hoax was provided in 1836 by Edgar Allen Poe, who wrote an essay entitled "Maelzel's Chess Player," outlining his suspicions about its operation:

> There is a man, Schlumberger . . . about the medium size, and has a remarkable stoop in the shoulders. Whether he professes to play chess or not, we are not informed. It is quite certain, however, that he is never to be seen during the exhibition of the Chess-Player, although frequently visible just before and after the exhibition.

Since that time, of course, technology has proceeded apace. With modern computers, the idea of a chess-playing machine is no longer outlandish, and there are even programs which have been awarded master ratings by the United States Chess Federation.* For the most part, these programs employ a relatively simple *lookahead* approach, which is applicable to a wide variety of games.

In many respects, playing a game of strategy is similar to solving a maze. Each move constitutes a choice point that leads to a different game path. Some of those paths lead to victory, others to defeat. A program that attempts to play such games intelligently must explore each of those paths to discover which options offer the best chances for victory.

Although games like chess and checkers may be more exciting, it is easier to illustrate the general structure of game-playing programs in the context of a much simpler game. For this purpose, one of the best examples is the game of Nim, which is simple enough to analyze in detail without being completely trivial.

The name Nim is derived from the Middle English verb *nimen* (meaning "to take" or "to steal") and applies to a variety of "take-away" games. One of the most widely known versions is 3-4-5 Nim, which is played with twelve pennies arranged to form three horizontal rows as shown below.

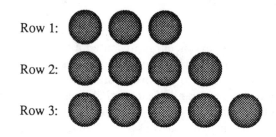

Playing alternately, each of the two players then takes away any number of pennies from any one of the horizontal rows. The object of the game is to take the last coin.

For example, the first player (player A) might take away all five pennies from the last row, leaving the position

---

*For a report on the Fourth World Computer Chess Championship and a commentary on the current state of chess programming, see the August 1984 issue of *Communications of the ACM*.

Player B must then remove pennies from one of the first two rows. Thus, player B might take three coins from the second row, leaving the position

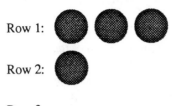

Row 1:

Row 2:

Row 3:

Looking at this position, it is not hard to see that the best move for player A is to take two of the coins in the first row. This results in the situation

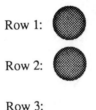

Row 1:

Row 2:

Row 3:

From here, player B must take one of the two remaining coins, and player A wins by taking the other.

Even though player A eventually discovered a winning move in the position 3-1-0 (that is, the position with three pennies in the first row, one in the second, and none in the third), each player has already thrown away a winning opportunity. Starting with the 3-4-5 arrangement, the first player can always win by making the right sequence of moves. After player A takes all five coins from the last row, however, player B had an opportunity to win the game by making the correct response. Before reading further, try to see if you can determine the mistake that each player made and devise a winning strategy.

After playing a few games, it becomes clear that some positions are much more promising than others. Using examples from the earlier game, the position 1-1-0 is fatal if it happens to be your move. You must take one of the two coins, leaving your opponent free to take the other. On the other hand, the configuration 3-1-0 is a "good" position from which to move, because taking two from the first row leaves your *opponent* in the 1-1-0 position. Since this position will guarantee a loss for your opponent, this implies a win for you.

The discovery that Nim has "good" and "bad" positions makes it possible to derive an algorithmic strategy for Nim, since it allows us to differentiate between "good" and "bad" moves. Intuitively, a move is "good" if it leaves your opponent with a "bad" position. Similarly, a position is "bad" if there are no "good" moves.

Although it may seem rather astonishing at first, this definition provides almost everything we need to write a program to play a winning game of Nim. The key here is to take the informal definitions of "good move" and "bad position" and turn those into recursive routines. Just as these concepts are each defined in terms of the other, each of these two routines will call the other in order to complete its operation.

Before we turn to the details of the implementation, it is helpful to define the data structures involved. First, we must be able to represent the current position in the game. A position in Nim is completely defined by the number of pennies in each row, which means that an array of three integers can be used to represent that position:

> **TYPE**
> **POSITION = ARRAY [1..3] OF INTEGER;**

A move is specified using two integers, the number of pennies taken (**N**) and the row from which those pennies are removed (**R**).

Given these definitions, we need to write two routines:

1. A function **BADPOSITION(POS)** which analyzes the position **POS** and returns **TRUE** if that position is a losing one, such as 1-1-0.
2. A procedure **FINDGOODMOVE(POS, N, R)** which, given the position **POS**, sets the parameters **N** and **R** so that they correspond to a good move. In some cases, however, there are no good moves, and **N** is set to zero to indicate that fact.

Each of the routines above operates by calling the other, as suggested by the informal definition. Thus, since a bad position is one in which there are no good moves, the function **BADPOSITION** simply calls **FINDGOODMOVE** and tests whether it returns 0 for N. Inside **FINDGOODMOVE**, we simply make every possible move until we find one that **BADPOSITION** tells us would cause our opponent to lose.

As of yet, we are not quite ready to write the actual code, since we have not defined a simple case which allows the recursive process to terminate. When the structure involves mutually recursive routines, the simple case test can usually go in any one of the routines, but the programmer must take care to ensure that the simple cases are always encountered somewhere along the line. Here, the most straightforward solution is to define the simplest "bad" positions explicitly as part of the **BADPOSITION** function.

At first glance, it seems that the position 1-1-0 (along with the symmetric positions 1-0-1 and 0-1-1) are the simplest bad positions. These, however, are not the best choices in this case. It is certainly true that these positions are so bad that we must lose no matter how our opponent plays, but things could be worse—we could have already lost. This is true if we reach the position 0-0-0, since our opponent must have taken the last penny.

Using the 0-0-0 case as the "fundamentally" bad position, we can write the code for the **BADPOSITION** function itself:

```
FUNCTION BADPOSITION(POS : POSITION) : BOOLEAN;
   VAR
      N, R : INTEGER;
   BEGIN
      IF POS[1] + POS[2] + POS[3] = 0 THEN
         BADPOSITION := TRUE
      ELSE
         BEGIN
            FINDGOODMOVE(POS, N, R);
            BADPOSITION := (N = 0)
         END
   END;
```

The coding of **FINDGOODMOVE** is somewhat trickier, mostly because Pascal makes it difficult to represent the necessary control structure. Intuitively, we would like to use the following definition:

```
PROCEDURE FINDGOODMOVE(POS : POSITION;
                       VAR N, R : INTEGER);
   BEGIN
      FOR R := 1 TO 3 DO
         FOR N := 1 TO POS[R] DO
            BEGIN
               POS[R] := POS[R] - N;
               IF BADPOSITION(POS) THEN return;
               POS[R] := POS[R] + N
            END;
      N := 0
   END;
```

The two **FOR** loops run through every possible move by selecting each row **R** and each legal value of **N** within that row. The body of the inner **FOR** loop makes each move by subtracting **N** from the Rth row. If **BADPOSITION** indicates that this new position would result in a loss for our opponent, we simply return from the procedure with the current values of **N** and **R** indicating the proper move. If not, we must *retract* the move we just tested with the statement

$$POS[R] := POS[R] + N$$

and try the next. If we get all the way to the end of each **FOR** loop without discovering a good move, **N** is set to zero to signify this condition and the program returns normally.

Unfortunately, Pascal forces a more complex coding because (1) the *return* statement does not exist, and (2) many compilers will not allow **VAR** parameters to be used as an index variable in a **FOR** loop. To get around these problems, each **FOR** loop must be rewritten as an equivalent **WHILE** loop that performs all of the index calculation explicitly. In addition, each of the new **WHILE** loops must also test a Boolean variable (called **FOUND** in this example) so that the loops will exit when the first good move is found. Making these changes results in the following code:

```
PROCEDURE FINDGOODMOVE(POS : POSITION;
                                  VAR N, R : INTEGER);
    VAR
      FOUND : BOOLEAN;
    BEGIN
      FOUND := FALSE;
      R := 0;
      WHILE (NOT FOUND) AND (R < 3) DO
        BEGIN
          R := R + 1;
          N := 0;
          WHILE (NOT FOUND) AND (N < POS[R]) DO
            BEGIN
              N := N + 1;
              POS[R] := POS[R] - N;
              IF BADPOSITION(POS) THEN FOUND := TRUE;
              POS[R] := POS[R] + N
            END
        END;
      IF NOT FOUND THEN N := 0
    END;
```

There is, however, one more Pascal wrinkle which we must consider. In Pascal, a procedure must ordinarily be defined before it is called. Here, **BAD-POSITION** calls **FINDGOODMOVE** and vice versa. This makes it impossible to follow the standard Pascal convention, since we can hardly define each of them before the other. To get around this problem, we must write a *forward declaration* for one of these procedures and then defer the actual definition of that procedure until after the other routine. Thus, if we choose to present these routines in the order in which they were developed, we must declare **FIND-GOODMOVE** as **FORWARD** using the lines

```
    PROCEDURE FINDGOODMOVE(POS : POSITION;
                           VAR N, R : INTEGER);
    FORWARD;
```

Following this, we write the definition for **BADPOSITION** and, finally, the actual code for **FINDGOODMOVE**.

When the code for **FINDGOODMOVE** is defined, Pascal insists that only the name of the procedure be specified and not the parameters. This makes the code much more difficult to follow, and it helps considerably to repeat the parameters as a comment to the procedure definition when it finally appears. Thus, the definition line of **FINDGOODMOVE** should probably be written as

```
    PROCEDURE FINDGOODMOVE(* POS:POSITION;
                           VAR N,R:INTEGER *);
```

followed by the actual code for the procedure.

The technique used to solve the Nim game is applicable to most two-handed strategy games, at least in theory. The difficulty with this approach is that most games are so complex that it is impossible to carry through with this analysis to any significant depth. For any game of reasonable complexity, a program is forced to limit the number of positions that it analyzes before making its turn.

In chess, for example, a player ordinarily has approximately 32 options at each turn. For each of these, the opponent has 32 possible replies, giving a rough total of 1000 positions after a single pair of moves. Since the average chess game involves about 40 moves for each player, the total number of different lines of play is therefore

$$1000^{40} = 10^{120}$$

This number is so preposterously large that, if a trillion computers on each of a trillion different planets were each able to analyze a trillion positions every second, it would still require more than

$$10^{75} \text{ years}$$

just to make the opening move!

In light of the enormous number of possibilities in a complete game, it is usually necessary to reduce this complexity by limiting the number of moves the program looks ahead and by eliminating unattractive variations relatively early.

Bibliographic Notes

Since artificial intelligence has acquired a significant popular appeal in recent years, there are many additional sources that describe the field as a whole and the algorithms that are used within it. Winston [1984] is an excellent text and describes many of the more important projects that have been undertaken in the field. For a more philosophical view of artificial intelligence, see Hofstadter and Dennett [1981], Weizenbaum [1976], and Turing [1947].

For further discussions of game playing by computer and strategies for improving algorithmic performance, see Shannon [1950], Berliner [1978], and Dewdney [1984].

Exercises

8-1. In the maze-solving program presented in this chapter, the program un-marks each square as it retreats during its backtracking phase. Since this removes the marks from the blind alleys that the program tries along the way, this process ensures that the final path will be displayed correctly. However, if the goal is to find the finish square in the shortest possible time, it is more efficient to leave these markers in place.

Discuss how this change affects the efficiency of the algorithm. In what situations would this new approach improve the program's performance?

8-2. Instead of awarding the game to the player who takes the last coin, Nim can also be played "in reverse" so that the player who takes the last coin *loses*. Redesign the strategy routines so that they give the correct moves for this "Reverse Nim" game, making as few changes to the routines as you can.

8-3. In the implementation of Nim used in the text, the simple case test included in **BADPOSITION** is not strictly necessary, and the function may be written simply as

```
FUNCTION BADPOSITION(VAR POS : POSITION) : BOOLEAN;
VAR
   N, R : INTEGER;
BEGIN
   FINDGOODMOVE(POS, N, R);
   BADPOSITION := (N = 0)
END;
```

Why does this work here?

8-4. The algorithm presented here to solve the Nim game is quite inefficient because it must analyze the same position many different times. For example, if the computer plays first from the 3-4-5 position, making the first move requires 25,771 calls to **BADPOSITION.** Since the total number of possible positions is only 120, this seems rather excessive.

The efficiency of the solution can be improved dramatically by having the program keep track of the result each time a position is analyzed. In this case, for example, we can construct a table with one entry for each of the 120 positions so that each entry indicates whether that position is "good," "bad," or "undetermined." Whenever we analyze a particular position in the course of the game, we can store this value in the table. The next time **BADPOSITION** is called, we can simply return the remembered information. Adopting this approach reduces the number of **BADPOSITION** calls to 213—improving the algorithmic performance by a factor greater than 100.

Write a complete program for the Nim game, using this extension. Try your program with longer initial rows. How quickly does it operate?

8-5. As presented, the recursive strategy discussed in the text is not directly applicable to games in which draws are possible, since one can no longer classify all positions as good or bad. On the other hand, it is reasonably simple to adapt this approach to handle draws by replacing **BADPOSITION** with a **RATEPOSITION** function that examines a position and assigns it a numerical score. For example, a simple assignment would be to use -1 to represent a losing position, $+1$ to represent a winning position, and 0 to represent one which is drawn.

With more than two possible ratings for positions, we need to define the recursive structure of the problem a little more carefully, although the basic structure remains the same. As before, the recursive insight is that the move that is best for us is the one that leaves our opponent in the worst position. Furthermore, the worst position is the one which has the least promising best move. Since our goal is to select a move which minimizes our opponent's best opportunity, this strategy is called the *minimax* algorithm.

Using this expanded strategy, write a program which plays a perfect game of Tic-Tac-Toe. In your implementation, the human player should make the first move, and the program should make the optimum reply.

8-6. Before Rubik made all similar puzzles seem rather pointless by comparison, many toy departments carried a puzzle with four painted cubes, marketed under the name of "Instant Insanity." The cube faces were painted with the colors green, white, blue, and red (symbolized as "G", "W", "B," and "R" in the diagrams), and each of the four cubes was composed of a distinct arrangement of colors. Unfolding the cubes makes it possible to show the color patterns.

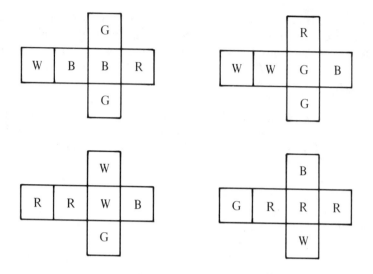

The object of the puzzle was to arrange these cubes into a line so that each line of four faces contained one face of each color.

Write a program to solve "Instant Insanity" by backtracking.

8-7. In the game of chess, most of the pieces move in relatively conventional patterns along horizontal, vertical, or diagonal lines. The knight, however, is an exception to this conservatism and moves in a rather unusual way. From its starting point, the knight moves two squares in any direction, horizontally or vertically, and then moves one square at a right angle to the original direction of motion. Thus, pictorially, the knight shown below can move to any of the eight black squares indicated by a diamond:

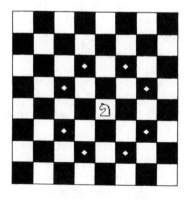

Even though its motion seems rather limited, the knight certainly gets around and can reach all 64 squares on the chessboard in as many moves.

A series of knight moves that traverses an entire chessboard without ever landing on the same square twice is known as a *knight's tour*. For example, the following diagram shows a knight's tour that returns to its initial square on the sixty-fourth move:

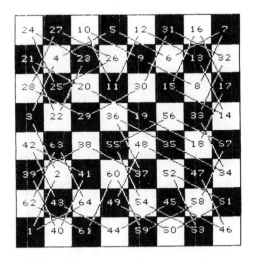

Write a Pascal program to find a knight's tour on an 8 × 8 chessboard, using a backtracking algorithm similar to the one used in the **MAZE** program.

Although the simple backtracking algorithm used to solve the maze in this chapter is sufficient to solve the knight's tour problem, it consumes far too much computation time to be practicable. Therefore, if you intend to test this program, you must improve the algorithmic performance substantially. One approach is to use a separate matrix to keep track of the number of squares from which any particular square can be reached. If making any move results in isolating some other square, then no solution is possible, and the program can immediately reject this possibility.

Graphical
Applications

Dost thou love pictures?
—*William Shakespeare*, The Taming of the Shrew

In teaching almost any discipline, one quickly comes to appreciate the truth of the old saying that "a picture is worth a thousand words." As a means of expressing a general concept, a simple diagram often conveys far more than many pages of descriptive text. In computer science, however, this adage has another interpretation—a program which generates a picture is often a thousand times more captivating and exciting than one whose output is limited to words or numbers.

There are many applications in computer graphics in which recursion plays an important role. In Chapter 1, one of the first examples used to illustrate the idea of recursion was the problem of generating "computer art" reminiscent of Mondrian's style:

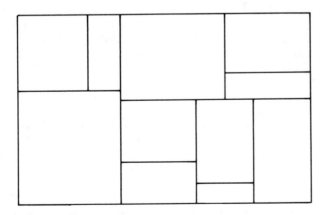

In this example, the complete picture is produced by dividing the original rectangle into parts which are, in turn, recursively subdivided until the program determines that the units are small enough to remain undivided.

9-1   Computer Graphics in Pascal

Writing a program to generate designs of this sort requires more than an understanding of its recursive structure. To see these programs work and thereby experience some of the excitement of computer graphics, it is necessary to have hardware and software capable of generating line drawings. Unfortunately, not all computer systems are equipped with a plotter or graphical display screen, and the examples in this chapter may not be implementable on all systems.

Even on those systems which have the appropriate hardware, using Pascal to generate diagrams is complicated by the fact that there is no standardized way to express graphical operations in the language. For the purposes of this chapter, we will assume the existence of two procedures, **SETPOINT** and **VECTOR,** which comprise our graphics library.

To understand the operation of these routines, imagine that you are using a pencil on a large piece of graph paper marked off in inches, with position (0,0) in the lower left-hand corner. A call to **SETPOINT(X,Y)** corresponds to picking up the pencil and moving it to a new position, indicated by its **X** and **Y** coordinates. A call to **VECTOR(DX,DY)** corresponds to drawing with the pencil from that point. The parameters **DX** and **DY** indicate the change in each coordinate. For example, the call

VECTOR(3,1)

indicates a line segment to the point 3 inches right and 1 inch up from its previous position.

If another call to **VECTOR** is made, the new line starts at the end of the last vector, so that a sequence of connected line segments can be drawn with a sequence of **VECTOR** calls. Calls to **SETPOINT** are required only to (1) set the initial position and (2) "lift your pencil" and move it to some other point on the page.

As an example, the rectangle

is drawn by the following sequence of Pascal statements:

> SETPOINT(2, 2);
> VECTOR(4, 0);
> VECTOR(0, 1);
> VECTOR(−4, 0);
> VECTOR(0, −1);

The first **SETPOINT** call sets the initial position to the point (2, 2) relative to the lower left-hand corner of the picture. The four **VECTOR** calls draw the bottom, right side, top, and left side of the box, in that order.

For many of the diagrams that appear in this chapter, a line segment will not be defined in terms of **X** and **Y** components, but by specifying its length and angular direction. For example, the line

is a 1-inch line inclined at an angle of 45°. To simplify diagrams of this sort, we will introduce the additional procedure **POLARVEC(R, THETA)**, which draws a line of length **R** moving in direction **THETA**, where **THETA** is an angle measured in degrees from the horizontal axis. Thus, the above line can be generated by calling

POLARVEC(1, 45)

The name **POLARVEC** is derived from the mathematical concept of "polar coordinates" and is based on the simple trigonometric relationships illustrated by the diagram below:

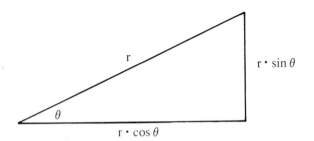

Using these relationships, it is easy to define **POLARVEC** in terms of **VECTOR** as follows:

```
PROCEDURE POLARVEC(R, THETA : REAL);
VAR
    RADIANS : REAL;
BEGIN
    RADIANS := THETA * PI/180;
    VECTOR(R * COS(RADIANS), R * SIN(RADIANS))
END;
```

The first statement in the procedure is used to convert the angle **THETA** from degrees into radians, since the **COS** and **SIN** routines expect angles to be expressed in this form.

9-2   Fractal Geometry

In his essay "On Poetry—A Rhapsody" written in 1735, Jonathan Swift offers the following observation on the continuity of natural processes:

> So, naturalists observe, a flea
> Hath smaller fleas that on him prey;
> And these have smaller still to bite 'em
> And so proceed, *ad infinitum.*

Taking a large amount of anatomical license, we can begin to diagram this process by representing an idealized flea as an equilateral triangle, shown here under considerable magnification:

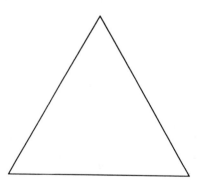

Putting a smaller flea on each of the two segments that form the back of the large flea gives the next stage in Swift's progression:

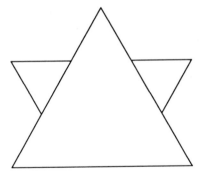

Each of these two new fleas plays host to two smaller fleas, which gives us the following menagerie:

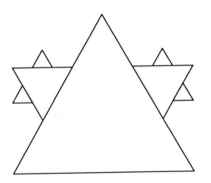

Clearly, this process can be repeated (as Swift suggests) *ad infinitum* with progressively smaller fleas.

Patterns such as this one turn out to have considerable mathematical significance and are known collectively as *fractals*. The term "fractal" was coined in 1975 by Benoit Mandelbrot, who synthesized several earlier mathematical discoveries to create a new geometry of fractal patterns. The principal defining characteristic of a fractal is that the components of a figure retain the same shape regardless of their size. In the case of the flea diagram, for example, each flea has exactly the same shape and, moreover, is oriented in exactly the same way, relative to the larger flea on which it sits. As it turns out, this property of "independence of scale" and the characteristic roughness of shape make fractals an excellent model for many of nature's irregularities.

The classic example used to illustrate the fractal tendencies of nature is the problem of determining the length of a coastline. The answer to this question depends on how closely you look. Conventional satellite photography reveals comparatively little detail, and any measurement based on those photographs will fail to account for peninsulas and inlets which are too small to be seen at that scale. With higher resolution, certain features may become visible, but there will still be smaller aberrations in the coastline (peninsulas on peninsulas in much the same manner as Swift's fleas) that escape our notice.

Before we investigate the relationship between fractals and recursive programming, it will be helpful to modify our "nested fleas" example to produce a figure which is somewhat more uniform in its structure. Once again, we will start with an equilateral triangle:

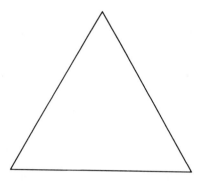

This figure represents the initial stage in the construction of a pattern called the Koch fractal, after the mathematician Helge von Koch. Since we have yet to add any fractal distortion to this pattern, we will call this triangle the Koch fractal of *order* 0.

To reach the next level, we put smaller triangles on all three sides and eliminate the interior lines:

This is the Koch fractal of order 1.

To obtain the next stage, we replace the center of *each* line segment with a smaller triangular wedge, resulting in the Koch fractal of order 2:

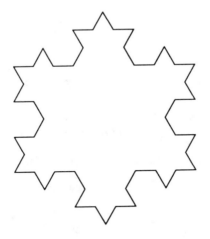

Repeating this process one more time produces the "snowflake" curve of order 3:

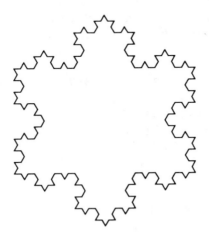

Writing a program to construct this pattern turns out to be surprisingly simple. As a start, we can certainly write a program to draw the original triangle:

```
PROCEDURE TRIANGLE(L : REAL);
  BEGIN
    SETPOINT(XCENTER − L/2, YCENTER − SQRT(3)*L/4);
    POLARVEC(L, 0);
    POLARVEC(L, 120);
    POLARVEC(L, 240)
  END;
```

In this program, **XCENTER** and **YCENTER** are constants which give the co-ordinate values of the center of the screen. The −L/2 and −SQRT(3)*L/4 in the **SETPOINT** call are derived from the trigonometric properties of an equilateral triangle and serve only to center the triangle on the screen. The important statements in the program are the **POLARVEC** calls which draw the triangle itself.

To reach the next level, each of these straight lines must be replaced with a "fractured" line of the following form:

Starting in the direction of the original line, we (1) draw a line one-third of the way across, (2) turn 60° to the right and draw another line of this same distance, (3) turn back so that we are pointing 60° degrees to the left of the original direction and draw another line, and (4) complete the figure with a line segment in the original direction. Thus, we can draw the order 1 fractal with code:

```
PROCEDURE FRACTLINE(L, THETA : REAL);
  BEGIN
    POLARVEC(L/3, THETA);
    POLARVEC(L/3, THETA − 60);
    POLARVEC(L/3, THETA + 60);
    POLARVEC(L/3, THETA)
  END;

PROCEDURE ORDER1FRACT(L : REAL);
  BEGIN
    SETPOINT(XCENTER − L/2, YCENTER − SQRT(3)*L/4);
    FRACTLINE(L, 0);
    FRACTLINE(L, 120);
    FRACTLINE(L, 240)
  END;
```

Of course, we could continue this process by adding additional nested procedures, but this is self-defeating. What we would like is a single mechanism that will allow the construction of a fractal of any order.

To do this, we need to recognize two things. Inside **FRACTLINE**, each of the calls to **POLARVEC** should, in the more general case, draw fractal lines of exactly the same form, but of the next smaller order. Thus, to draw a fractal line of order 3, we simply draw four lines of order 2 in the appropriate directions. The other observation is that the order 0 line has no distortions and is simply a straight line. This leads immediately to the complete recursive solution.

```
PROCEDURE FRACTLINE(ORDER : INTEGER; L, THETA : REAL);
BEGIN
   IF ORDER = 0 THEN
      POLARVEC(L, THETA)
   ELSE
   BEGIN
      FRACTLINE(ORDER-1, L/3, THETA);
      FRACTLINE(ORDER-1, L/3, THETA - 60);
      FRACTLINE(ORDER-1, L/3, THETA + 60);
      FRACTLINE(ORDER-1, L/3, THETA)
   END
END;

PROCEDURE SNOWFLAKE(ORDER : INTEGER; L : REAL);
BEGIN
   SETPOINT(XCENTER - L/2, YCENTER - SQRT(3)*L/4);
   FRACTLINE(ORDER, L, 0);
   FRACTLINE(ORDER, L, 120);
   FRACTLINE(ORDER, L, 240)
END;
```

Bibliographic Notes

The use of computer graphics has gained considerable momentum in recent years with the success of the personal computer, many of which offer some form of graphical capabilities. In that area, one of the more delightful applications has been in teaching elementary programming concepts by providing a simple programming environment based on manipulating some entity on a screen. The most well-known project of this sort is the Project LOGO Turtle, which is described in Papert [1981] and Abelson and diSessa [1981]. Fractal patterns are described in Mandlebrot [1983]. For a more advanced treatment of computer graphics, consult Foley and Van Dam [1982].

Exercises

9.1   "Once upon a time, there was a sensible straight line who was hopelessly in love with a dot."

With these words, we are introduced to the hero of Norton Juster's delightful story *The Dot and the Line*, who eventually discovers the versatility of his linear form and lives "if not happily ever after, at least reasonably so." In demonstrating his new found talents to the object of his affection, one of his displays has the following form:

ENIGMATIC

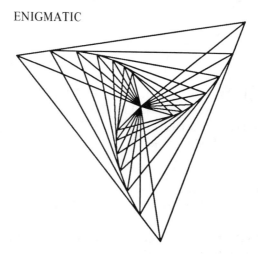

Taking advantage of the fact that this diagram consists of several repetitions of the figure

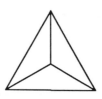

drawn at various sizes and angles, write a Pascal program to generate this diagram. In this example, recursion is not strictly necessary, but it provides a fairly simple way of generating the repeated figures.

9-2. Using the definitions of **SETPOINT** and **VECTOR** given in this chapter, complete the code for the Cubist painting program described in Chapter 1. The difficulty here lies in choosing some criterion for terminating the recursion. Ideally, the decision of whether to split a subcanvas should depend on the size of the canvas in some way; larger problems should be subdivided more often than smaller ones.

9-3. Modify the **SNOWFLAKE** program to generate the flea diagrams used in this text. Although the figures are quite similar in structure, the need to retrace interior lines adds some interesting complexity.

9-4. In the text, the scale-independence of fractals is described in terms of the problem of coastline measurement. Fractals can also be used to generate random coastal patterns in a way that illustrates their application to natural phenomena.

For example, suppose that we wanted to generate a random stretch of coastline, linking two points A and B. The unimaginative solution (using creativity of order 0) is to connect them with a simple straight line.

To give this coastline some semblance of irregularity, we can introduce triangular fractal distortion exactly like that used to generate the snowflake design. The difference here is that the direction in which the triangle points is chosen at random, so that some points go "up" while others go "down." Thus, after two levels of fractal decomposition, we might get the line

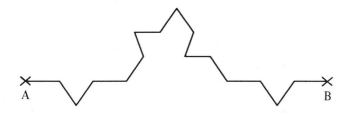

Repeating this with a fractal of order 5 produces a fairly elaborate coastline which shows almost no trace of the triangular pattern used to generate it:

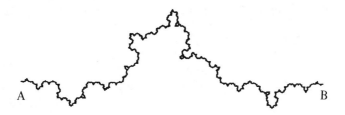

Using the method described here, write a program that generates a fractal coastline for any given order.

9-5. One particularly interesting class of fractals can be constructed using the following technique. Start with any polygon, such as the pentagon shown below:

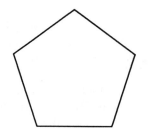

At each corner of the figure, draw an identical figure one-half the size, outside of the original figure. This creates the following structure:

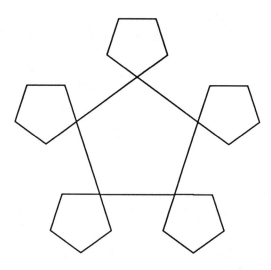

Using recursion, repeat this process for each of the newly generated corners to any desired depth. Thus the next figure in this process would be:

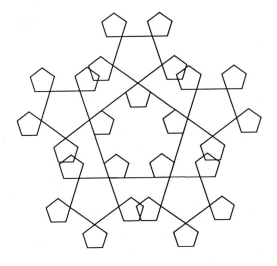

Write a Pascal program which generates diagrams that follow this rule. In designing your program, you should make it easy to change any of the following:

(a)   The number of sides in the basic figure
(b)   The length of the edge
(c)   The number of recursive levels

Recursive Data

If somebody there chanced to be
Who loved me in a manner true
My heart would point him out to me
And I would point him out to you.
                        —*Gilbert and Sullivan,* Ruddigore

One of the important strengths of Pascal as a programming language lies in the flexibility it offers for data representation. Many languages, including FOR-TRAN, restrict the user to certain simple objects and arrays. In Pascal, the programmer has considerably more freedom, primarily because the language supports two important mechanisms for defining new data structures: *records* and *pointers*. A record makes it possible to combine related data values so that the collection may be treated as a single object. Pointers provide a powerful mechanism for expressing the relationships between data objects and are quite efficient when those relationships are subject to frequent change.

Inasmuch as the focus of this book is on recursion and not on the mechanics of structured data, this chapter assumes that the reader has some familiarity with records and pointers and with the use of these structures in Pascal. In this chapter, our purpose is to show how these pointers, particularly when used in conjunction with record objects, make it possible to extend the general idea of recursion from algorithms into the domain of data structures.

In Pascal, a recursive data type is one in which the complete description of that type includes a pointer to that type. For example, the type definition

> **TYPE**
> **PTRPTR = ↑ PTRPTR;**

is recursive, since it defines the type **PTRPTR** as a pointer to objects of that type. Such a definition, however, is at most marginally useful (see Exercise 10-4), since it contains no data other than the pointer itself.

A more useful situation occurs when the recursive type consists of a record in which certain fields contain data values and others are pointers used to link individual objects together to form more complicated structures. For example, the definition

```
TYPE
   OBJECT = RECORD
                     DATA : INTEGER;
                     LINK : ↑OBJECT
            END;
```

introduces the new data type **OBJECT** which consists of a **DATA** field holding an integer and a **LINK** field of type "pointer to an **OBJECT**." Often, the **LINK** field will be used to point to another **OBJECT** to establish a chained relationship between the two. At the end of such chains, however, the **LINK** field will have the special value **NIL,** which is a legal value for any pointer type. Each of these possibilities is illustrated in the diagram below which shows a *linked list* consisting of the integers 1 and 2.

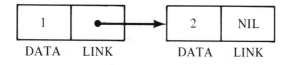

In many cases, it will be more convenient to consider the pointer as the principal data entity rather than the record type itself. In this case, we will name the pointer type first and subsequently define the record type which contains it. Thus, this definition will more commonly be expressed using two **TYPE** section entries:

```
TYPE
   LIST = ↑OBJECT;

   OBJECT = RECORD
                     DATA : INTEGER;
                     LINK : LIST
            END;
```

When recursive data structures are introduced in conventional textbooks, the algorithms to manipulate those structures are often presented in an iterative form. In many cases, this sacrifices the symmetry and simplicity that are possible in a recursive solution. Whenever a recursive data structure is used, the relationships implied by that structure provide a natural model for the recursive algorithms used to work with that data. In the examples which follow, the emphasis is placed on the recursive character of the solutions and the relationship between the data and algorithmic structure.

10-1   Representing Strings as Linked Lists

Despite the power provided by Pascal's data structure facility, there are also several limitations. One of the most serious problems arises in the handling of character strings. In standard Pascal, all string variables are declared as character arrays of a specific length. Once declared, that length is not allowed to vary. For those applications which specify specific field widths for all data items (as in the case, for example, of a record where the name field fits in columns 1 to 25, the address in 26 to 50, and so forth), this is of little concern. However, if the application must work with strings of different lengths, this becomes a serious problem.

One approach to the representation of variable length strings is to use a linked list of characters instead of an array. To establish such a structure, we need the following data definitions:

```
TYPE
  STR  =  ↑CELL;
  CELL  =  RECORD
               CH    : CHAR;
               LINK : STR
           END;
```

In this implementation, if the string 'ABC' is stored in the variable **S**, we can diagram the internal structure as follows:

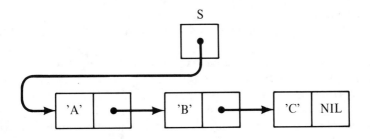

Similarly, if **T** contains only the single character 'D', it has the form:

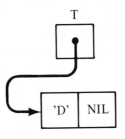

Finally, if **Z** contains the empty string (the string with no characters at all), it is represented as:

It is important to recognize that the variables **S, T,** and **Z** in these examples are declared as objects of type **STR,** and that the object referred to by that name consists only of the *pointer* itself. The characters that form the string can easily be determined by following the chain, but the **STR** structure itself is a small object of a fixed size.

In addition to the definition of **STR,** a general string-handling package also requires a set of subroutines that operate on string data represented in that form. From our experience with strings, we know that there are certain basic operations that must be implemented in order to make strings behave in an intuitively satisfying way. Given a new data type, the functions and procedures which implement the necessary basic operations are called *primitives* for that type. In the case of strings, for example, we would certainly want the following primitives:

LENGTH(S)   Returns the length of **S**, where **S** is a **STR**.
WRITESTR(S)   Writes **S** to the standard output.
READSTR(S)   Reads **S** from the standard input.
CONCAT(S,T)   Concatenates **S** and **T** and returns the combined **STR**.

Before we write the code for each of these primitives, it will help to adopt a somewhat different perspective on the definition of a string. At the level of machine representation, a string is a pointer to a **CELL** which, in turn, is a record containing both a character and a pointer. At the level of the application, we must be able to think of a string as a single conceptual entity consisting of a sequence of characters, irrespective of the underlying representation. To develop the primitives, however, it is useful to adopt an intermediate view in which the definition of string (1) parallels the recursive structure of the representation and (2) makes it possible to think about that string at a relatively abstract conceptual level.

To do this, it helps to adopt an informal definition for a string which is reminiscent of the definition of a noun phrase from Chapter 4:

A string is either
  1. The empty string
  2. A character followed by a string

The principal advantage of this definition is that it provides a model for recursive routines which take strings as arguments. The simple case in such routines is always that of the empty string. In the general case, the recursive decomposition will consist of dividing a string into its first character and the remainder. For example, the function to calculate the length of a string can be defined as follows:

```
FUNCTION LENGTH(S : STR) : INTEGER;
   BEGIN
      IF S = NIL THEN
         LENGTH := 0
      ELSE
         LENGTH := 1 + LENGTH(S ↑ .LINK)
   END;
```

Rendered in English, this function indicates that (1) the length of the empty string is zero and (2) that the length of any other string is one more than the length of that string after taking away its first character.

The procedure **WRITESTR** follows much the same form. To write the empty string, we do nothing at all. Any other string is written by (1) writing out its first character and (2) recursively writing out the rest of the string:

```
PROCEDURE WRITESTR(S : STR);
   BEGIN
      IF S <> NIL THEN
         BEGIN
            WRITE(S ↑ .CH);
            WRITESTR(S ↑ .LINK)
         END
   END;
```

The operations **READSTR** and **CONCAT,** however, are somewhat different from the two primitives described above. In each of these operations, it is necessary to create a new object of type **STR** to be returned to the calling program. Given our current model of string operations, it is not surprising to find that the operation of constructing a string is symmetrical to that of breaking it down. We decompose a string by removing its first character; we build up by adding new first characters to existing strings.

Since this operation of combining a character and a string to form a new string is common to both **READSTR** and **CONCAT** (as well as many of the other functions we are likely to define), it is useful to implement this operation separately as the **MAKESTR** function:

```
FUNCTION MAKESTR(C : CHAR; L : STR) : STR;
VAR
  RESULT : STR;
BEGIN
    NEW(RESULT);
    RESULT ↑ .CH := C;
    RESULT ↑ .LINK := L;
    MAKESTR := RESULT
END;
```

The first statement in this function obtains a new cell from the dynamic storage allocator and stores a pointer to that cell in the variable **RESULT**. The subsequent statements simply assign the parameters **C** and **L** to the appropriate components of that cell and return the pointer as the value of the function.

Once **MAKESTR** has been defined, we are in a position to define each of the remaining primitives. The definition of **READSTR** looks somewhat convoluted, but the complexity exists primarily to accomodate the vagaries of Pascal's input/output structure. In standard Pascal, **EOLN** is initially true, so that the **READLN** call must be performed before any characters are read. This is accomplished by defining the **READSTR** operation using two routines. The **READSTR** routine itself performs only the necessary setup operations, leaving the real work to the function **READS**.

```
FUNCTION READS : STR;
VAR
  CH : CHAR;
BEGIN
  IF EOLN THEN
    READS := NIL
  ELSE
    BEGIN
      READ(CH);
      READS := MAKESTR(CH, READS)
    END
END;
```

```
PROCEDURE READSTR(VAR S : STR);
BEGIN
  READLN;
  S := READS
END;
```

The definition of **CONCAT** is reasonably straightforward, now that **MAKESTR** is available as a tool.

```
FUNCTION CONCAT(X, Y : STR) : STR;
  BEGIN
  IF X = NIL THEN
    CONCAT := Y
  ELSE
    CONCAT := MAKESTR(X ↑ .CH, CONCAT(X ↑ .LINK, Y))
  END;
```

Go through this definition on your own to see exactly how its internal operation corresponds to the recursive definition.

10-2 Binary Trees

In Chapter 7, we looked at a variety of sorting algorithms and compared their algorithmic efficiencies. Although we discovered that certain algorithms are considerably more efficient than others, sorting continues to be a time-consuming operation when large amounts of data are involved. In many practical applications, it is advantageous to avoid these sorting operations by inserting each new data value in its proper sequence at the time it is first encountered.

Inserting a new item in its proper order can be broken down into two distinct operations. First, we must find the correct position for the new item. This constitutes the *search* phase of the operation. Second, we must be able to insert the new item in its proper place without disturbing the overall arrangement. This is the *insert* phase. In analyzing various strategies for performing these operations, we will discover that the efficiency of each phase depends quite dramatically on the data structure we use. Moreover, one data structure may be much more efficient for part of the operation and yet be preposterously inefficient for the other.

For example, storing the numbers in a large array provides for a relatively efficient search phase. As we discovered in Chapter 2, searching a sorted list for a particular element can be performed in log N time by using the binary search technique. On the other hand, the insert phase is extremely complicated here. To ensure that the array remains sorted, it is necessary to move the rest of the elements in the entire array over one position to make room for the new element. Since this could require moving every element if the new element occurs at the beginning, this phase of the operation requires order N steps.

By contrast, representing the array as a linked list has precisely the opposite computational properties. Here, once we have located the correct position, the insertion phase can be performed in constant time. All we need to do is allocate a new cell and adjust the internal pointers to maintain the order. Here, however, the search phase can no longer exploit the advantages of binary search, since there is no simple way to perform the operation of selecting the middle element.

Thus, for each of these representations, we have the following computational efficiencies:

	Search	Insert	Total
Sorted array	$O(\log N)$	$O(N)$	$O(N)$
Linked list	$O(N)$	$O(1)$	$O(N)$

Since each of these representations leads to an order N combined time for the search-and-insert operation, neither structure is ideal for the task of maintaining a sorted list. What we would like to find is some mechanism which combines the search properties of a sorted array with the insertion properties of a linked list. To do this, it helps to start with the linked list arrangement and repair its flaws.

Conceptually, if we were to store the numbers 1 through 7 in a linked list, we would use a structure of the following form:

Our current difficulty arises from the fact that, given this structure, there is no simple way to find the center of the list. To find the "4", we are obligated to start at the beginning and follow the chain through each of its links, until we reach that cell.

In seeking a solution to this problem, one idea is simply to change the list structure so that the initial pointer always indicates the center of the list rather than its beginning:

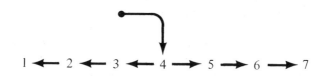

In making this change, it is also necessary to reverse the direction of the first three pointers so that the elements 1, 2, and 3 continue to be accessible. Notice that this requires the element 4 to be associated with *two* pointers: one points to a list of those elements that precede 4, and one points to those which follow it.

At this point, however, we have only solved the first level of the problem. In this diagram, we can certainly find the center of the original list, but we are back to our original problem for each of the sublists. The answer is simply to apply this same transformation recursively, so that each sublist begins in the middle and proceeds in both directions. Applying this to the above diagram results in the structure

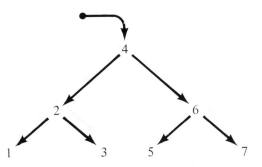

At each level, this structure branches to form two smaller structures, which are in turn decomposed in precisely the same way. In computer science, this type of structure is called a *binary tree*. In a binary tree such as the one represented here, each individual element is called a *node*. The topmost node in a tree is called the *root* of the tree, and the nodes at the bottom which have no descendants are called *terminal nodes*.*

In the binary tree represented here, each node is associated with an integer value and a pair of pointers indicating two subtrees: the subtree which contains all elements *before* the current value and one containing those elements *after* this node. This description leads directly to the appropriate data structure in Pascal:

```
TYPE
  TREE  =  ↑NODE;

  NODE = RECORD
           VALUE : INTEGER;
           BEFORE : TREE;
           AFTER : TREE
         END;
```

As indicated by the earlier discussion, our central task is to develop a mechanism to allow the incremental insertion of new data into a binary tree. Before we look at that aspect of the problem, however, it will help to design a simpler primitive for trees so that we can get a sense of how tree manipulation is performed.

---

*Since many programmers have rarely been seen outside during the daylight hours, it is perhaps understandable that the prevailing convention in computer science is to draw trees with their root at the top and their branches pointing down. Those offended by this affront to nature are invited to turn this book upside down before proceeding.

For example, it would be quite useful to have a procedure which prints out all the values in a tree in properly sorted order. Thus, given the tree

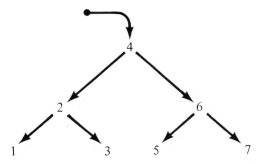

we need a procedure **WRITETREE** which prints the values

$$1 \quad 2 \quad 3 \quad 4 \quad 5 \quad 6 \quad 7$$

in that order. At first, this appears difficult, since this tree initially points to the 4, which must be printed in the middle of the sequence.

The solution to this problem lies in writing a procedure whose structure parallels that of the data. In working with strings earlier, we adopted a definition of "string" that made the recursive structure explicit. Here, we should adopt a similar definition for binary trees, and define a tree as follows:

A tree is either
    1. The empty tree represented by **NIL**
    2. A number preceded and followed by a tree

Working from this definition, the **WRITETREE** primitive has almost exactly the same structure as the **WRITESTR** primitive. Once again, there is nothing at all to do in the simple case. Otherwise, we have a number preceded and followed by sets of numbers represented as binary trees. To obtain a correctly ordered display, all we need do is write out the numbers in the "before" collection, write the current value, and then write out the numbers in the "after" collection, as indicated by the following code:

```
PROCEDURE WRITETREE(T : TREE);
BEGIN
  IF T <> NIL THEN
    BEGIN
      WRITETREE(T ↑ .BEFORE);
      WRITE(T ↑ .VALUE : 5);
      WRITETREE(T ↑ .AFTER)
    END
END;
```

From here, we are in a position to develop the procedure which performs the actual insertion. Here, we will define this operation to have the parameter structure

PROCEDURE INSERTTREE(N : INTEGER; VAR T : TREE);

The effect of this procedure is to insert the value **N** into its proper position in the tree **T**. Since the tree pointer will in some cases be changed, **T** is passed as a **VAR** parameter.

Like any recursive procedure, **INSERTTREE** must correctly handle the simple cases. Here, for example, if we start with an empty tree, the program must correctly create a new node and change the tree **T** so that it points to this new node. This is accomplished by the following code:

```
PROCEDURE INSERTTREE(N : INTEGER; VAR T : TREE);
  BEGIN
   IF T = NIL THEN
    BEGIN
      NEW(T);
      T↑.VALUE := N;
      T↑.BEFORE := NIL;
      T↑.AFTER := NIL
    END
   ELSE
    . . . recursive case . . .
  END;
```

Thus, if we start with an empty binary tree called **DATATREE,** calling **INSERTTREE(6, DATATREE)** will result in the following structure:

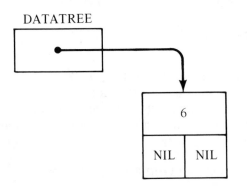

If we already have some nodes of the tree in place, the operation is somewhat more elaborate. In general, given any new value, we must first check it against the number in the topmost node of the tree. If it is less than the value in that position, we know that the new value should be entered in the T ↑ .BEFORE subtree. Similarly, a value larger than that in the current node should be entered in the T ↑ .AFTER subtree. Thus, we should be able to complete the problem by calling either

INSERTTREE(N, T ↑ .BEFORE)

or

INSERTTREE(N, T ↑ .AFTER)

We should also decide what action should be taken in the case of duplicate nodes. The problem description did not define what action should occur if the value 6 is to be entered into a tree which already contains a node labeled 6. Here, we will make the arbitrary assumption that each value is to be stored only once in the tree, and that requests which would duplicate a value are simply ignored. This gives rise to the complete definition of **INSERTTREE** shown below:

```
PROCEDURE INSERTTREE(N : INTEGER; VAR T : TREE);
BEGIN
  IF T = NIL THEN
    BEGIN
      NEW(T);
      T ↑ .VALUE := N;
      T ↑ .BEFORE := NIL;
      T ↑ .AFTER := NIL
    END
  ELSE
    IF T ↑ .VALUE <> N THEN
      IF T ↑ .VALUE > N THEN
        INSERTTREE(N, T ↑ .BEFORE)
      ELSE
        INSERTTREE(N, T ↑ .AFTER)
END;
```

To illustrate the operation of **INSERTTREE,** suppose that **DATATREE** consists of the single number 6 when **INSERTTREE(3, DATATREE)** is called. Since 3 is smaller than the value of the current node, the procedure recognizes that this value should be inserted in the "before" subtree and the procedure calls

INSERTTREE(N, T ↑ .BEFORE)

Since **T** is defined as a **VAR** parameter, this means that the next recursive level uses a value of **T** which is shared with the actual component of the record at this level. Thus, the parameter **T** in the second level call is precisely the cell indicated by the dotted line below:

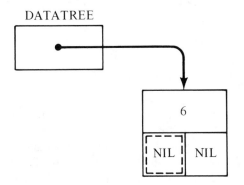

This means that when the new node is allocated and assigned to **T,** the result will be:

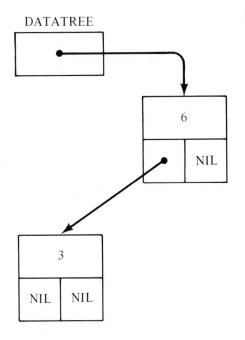

If from here we were to insert the value 5, the first level of the recursive call would choose the **BEFORE** subtree of the node containing 6, the second would choose the **AFTER** subtree of the node containing 3, and the new tree pointer would be inserted in the **AFTER** component of that node.

As long as the input values are distributed in a reasonably random order, inserting a new value into its proper position in a binary tree requires only log N time. This represents an enormous increase in computational efficiency over arrays and linked lists, both of which require linear time for the combined "find-and-insert" operation.

10-3   Expression Trees

In addition to offering a computationally efficient mechanism for maintaining a sorted collection of data, tree structures have many other applications in computer science. Of these, one of the most important arises in the design and implementation of compilers, which are responsible for translating statements in a programming language into a form more easily handled by the computer itself.

In order to perform this translation, compilers must translate the statements in the source language into an internal form that represents the semantic structure of the program. For example, given the Pascal expression

$$2 * X + 3$$

the compiler must convert it into an internal representation that reflects the order of operations. Here, since multiplication is an operator with greater precedence than addition, this statement is the same as

$$(2 * X) + 3$$

and the compiler must interpret this as a sum of two subexpressions: (a) the product of 2 and X and (b) the number 3. Conventionally, this is represented internally by the compiler in the form of a binary *expression tree:*

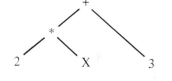

One of the advantages of tree representation is that the order of operations is explicit and there is no need to represent parentheses or precedence values within the tree itself. For example, even though the parentheses are significant in the expressions

$$(2 * X) + 3 \quad \text{and} \quad 2 * (X + 3)$$

the corresponding expression trees indicate the order of operations through the structure of the tree itself:

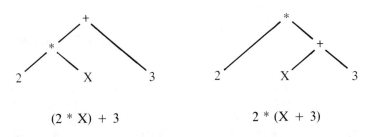

$$(2 * X) + 3 \qquad\qquad 2 * (X + 3)$$

Before we proceed to the internal representation, it is necessary to define the notion of expression in greater detail. Conceptually, we know that expressions consist of numbers, variables (limited in this example to single characters in the range A to Z), and some set of operators (here $+$, $-$, $*$, and $/$). Parentheses may be used in the original expression to indicate the order of operations, but they are not preserved in the internal representation.

In attempting to translate this conceptual structure into Pascal, however, we encounter a slightly different situation from that in our earlier use of binary trees. In the example from the previous section, all nodes in the tree had exactly the same form. Here, there are three different node types: *numbers, variables,* and *compound* nodes which indicate the application of some operator to two subexpressions. Each of these node types requires different internal data. More specifically, numbers must be associated with their value, variables with their name, and compound nodes must include an operator and each of the two subexpressions.

In order to represent structures in which the components have several possible forms, Pascal offers a special mechanism called *variant records*. In a variant record, one field is selected as the *tag field* and is used to determine the interpretation of the remaining fields. Here, for example, we have three different types of expression node. Depending on its type, the node itself contains an integer, a character, or the data required for a compound node.

The first step in establishing a variant record of this sort is to define a new scalar type which specifies the possible formats:

XTYPES = (NUMBER, VARIABLE, COMPOUND);

This statement defines the names **NUMBER, VARIABLE,** and **COMPOUND** as constants of the type **XTYPES** (represented internally by the integers 0, 1, and 2). In working with expression trees, we will use these values in their symbolic form to differentiate the three expression formats. Expressions themselves are defined by the following **TYPE** section entries:

EXP = ↑ EXPNODE;

EXPNODE =
  RECORD
    CASE EXPTYPE : XTYPES OF
      NUMBER : (VALUE : INTEGER);
      VARIABLE : (NAME : CHAR);
      COMPOUND : (OP : CHAR; LHS, RHS : EXP)
  END;

Once again, the **EXP** type is the *pointer,* and the record type itself will not ordinarily appear explicitly in the code. The record definition specifies the structure of a node in each of the three possible interpretations. According to the setting of the **EXPTYPE** field, an **EXPNODE** has either (1) an integer field named **VALUE,** (2) a character field called **NAME,** or (3) an **OP** field representing an operator character and two additional expression pointers indicating the left and right subexpressions.

To make the use of this structure more clear, the following diagram provides a pictorial representation of the expression

$$2 * X + 3$$

as it would appear internally:

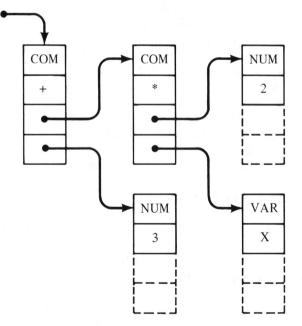

In many ways, the most interesting aspect of this problem is that of generating the expression tree from its original programming language form. This phase of the compilation process is known as *parsing* the input text. Unfortunately, an adequate treatment of parsing is beyond the scope of this book. On the other hand, many of the other operations that may be performed on expression trees are quite straightforward and fit well into the context of this discussion.

For example, one operation that might prove useful (particularly during the development and debugging of the compiler itself) is that of converting an expression tree back into its programming language form. This "unparsing" operation is considerably easier than that of parsing it in the first place and is accomplished by the following code:

```
PROCEDURE UNPARSE(T : EXP);
   BEGIN
      CASE T ↑ .EXPTYPE OF
         NUMBER    : WRITE(T ↑ .VALUE : 1);
         VARIABLE : WRITE(T ↑ .NAME);
         COMPOUND : BEGIN
                       WRITE('(');
                       UNPARSE(T ↑ .LHS);
                       WRITE(' ', T ↑ .OP, ' ');
                       UNPARSE(T ↑ .RHS);
                       WRITE(')')
                    END
      END
   END;
```

Given an expression tree **T**, the **UNPARSE** program looks at the expression type and handles it accordingly. Numbers and variables constitute the simple cases; compounds are handled by recursively unparsing each subexpression, separating them with the appropriate operator, and enclosing the entire unit in parentheses. Thus, given the expression tree

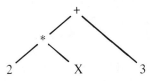

the **UNPARSE** procedure will produce

$$((2 * X) + 3)$$

Bibliographic Notes

This chapter contains material which is generally the central focus of a course
on data structures, and there are several good reference texts which will provide
further information, including Wirth [1976], Tenenbaum and Augenstein [1981],
and Reingold and Hansen [1983].

Exercises

10-1. Rewrite each of the functions **LENGTH, WRITESTR, READSTR** and
CONCAT so that they operate iteratively without using recursion.

10-2. Given a string S, what does the following procedure do?

```
PROCEDURE UNKNOWN(S : STR);
  BEGIN
    IF S <> NIL THEN
      BEGIN
        UNKNOWN(S ↑ .LINK);
        WRITE(S ↑ .CH)
      END
  END;
```

Can you duplicate its function without using recursion?

10-3. In addition to the string primitives defined in the chapter, there are several
others which might be useful in a general library of string routines. As
examples, implement the following:

INDEX(S, CH)	Given the string **S** and the character **CH,** returns the position number corresponding to the first occurrence of **CH** in that string. For example, if **S** is the **STR** representation of **'HELLO'**, and **CH** is **L**, **INDEX** should return 3. If **CH** does not occur in **L**, **INDEX** should return 0.
SEG(S,N1,N2)	Returns the substring of **S** that lies between character positions **N1** and **N2,** inclusive. Remember that this function must return an object of type **STR** and may not change the value of the original string **S**.
STREQ(S, T)	Returns **TRUE** if strings **S** and **T** contain an identical sequence of characters.
REVERSE(S)	Returns a string composed of the characters in **S** in reverse order.

Try implementing the **INDEX** function iteratively. Be sure you test your implementation thoroughly, since this is harder than it may appear.

10-4. [Mathematically oriented] Although the type **PTRPTR** defined by

$$\textbf{TYPE}$$
$$\textbf{PTRPTR} = \uparrow \textbf{PTRPTR};$$

seems rather useless, there are an infinite number of objects of this type. For example, **NIL** is certainly a **PTRPTR,** but so is a pointer to **NIL,** and a pointer to that, and so on.

Since this definition allows us to define chains of pointers with any arbitrary length, we can use this structure as a model of the natural numbers 0, 1, 2, 3, . . . by associating each number with the **PTRPTR** chain of that length. Thus, 0 is associated with **NIL,** 1 with a pointer to **NIL,** and so forth.

Using this structure, write the functions **ADD1** and **SUB1** which, given a number represented as a **PTRPTR,** return its successor and predecessor, respectively. Armed with these functions, write **ADD** and **MULT** for **PTRPTR**'s.

10-5. As presented here, using the type **STR** to represent string data is very inefficient in terms of space utilization. In this representation, each character requires enough storage to hold both a character and a pointer object which is often considerably larger than that character.

To reduce this overhead, it is possible to define strings so that the chaining is done on the basis of character blocks rather than individual characters. Design a data structure which accommodates more than a single character within each cell. Discuss the effect that this new representation has on the primitive functions, along with any design trade-offs you anticipate.

10-6. In the discussion of the complexity order of insertion using a binary tree, the statement includes the disclaimer that the input values must be distributed in a reasonably random order. What would happen if an overzealous clerk sorted a large batch of entries by hand and then fed them into the binary tree insertion mechanism in that order?

10-7. Diagram the tree that results if the following values are inserted into an empty tree in the specified order:

90, 99, 63, 82, 93, 76, 81, 76

10-8. The *depth* of a binary tree is defined to be the length of the longest path that runs from the root of the tree to one of the terminal nodes. Thus, in the tree

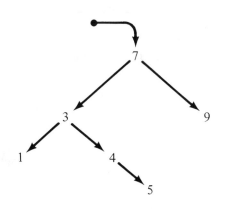

the depth of the tree is 3, since that is the length of the path from node 7 to node 5.

Write a Pascal function **DEPTH(T)** which, given a binary tree T, returns its depth.

10-9. Write a Pascal procedure **DELETETREE(N, T),** which deletes the node with value **N** from tree **T**. If there is no node with the indicated value, **DELETETREE** should have no effect.

10-10. Many of the electronic calculators produced in the early 1970s required the user to enter the data in *reverse Polish notation*. In this representation, the operands to an expression are written first, followed by the operator. For example, the expression

$$2 * X + 3$$

in conventional algebraic notation becomes

$$2 \ X * 3 +$$

in reverse Polish.

Using **UNPARSE** as a model, write the procedure **RPN** which prints an expression tree in reverse Polish form.

10-11. Many conversational language systems (including most implementations of BASIC) are based on an *interpreter* rather than a traditional compiler. In a compiler, the expression trees must be translated into machine language instructions which can then be executed by the machine. In an interpreter, the necessary operations are performed by working directly from the expression tree. For example, given the tree

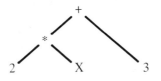

the program would look up the current value of **X**, multiply it by 2, add 3, and return the answer directly.

Suppose that the system includes a variable **SYMTAB** (for *symbol table*) which is simply an array of the current values for each of the variables **A** to **Z**:

SYMTAB : ARRAY ['A'..'Z'] OF INTEGER;

Using this, write a Pascal function **EVAL** which takes an expression tree and calculates its value by applying the appropriate operations to each of the terms making up the expression. Thus, if **SYMTAB['X']** contains 4, evaluating the above expression tree would return 11.

10-12. In the **UNPARSE** example given in the text, parentheses appear frequently in the output text, even if they were not present in the original text. In order to simplify the output form by eliminating unnecessary parentheses, **UNPARSE** must "know" about the precedence structure of the operators involved.

Extend the definition of **UNPARSE** so that it displays an expression tree using as few parentheses as possible. To do this will require splitting **UNPARSE** into several internal routines.

10-13. In light of its enormous importance in the design of compilers, the problem of parsing arithmetic expressions is one of the most thoroughly studied areas in computer science. Unfortunately, constructing an expression parser that takes advantage of the results of that study requires the presentation of so much theory as to be beyond the scope of this book.

There are, however, other types of expressions for which the design of a parser is feasible. One of these comes from the game WFF'N PROOF, designed by Layman E. Allen of the University of Michigan. WFF'N PROOF is based on symbolic logic and involves recognizing and manipulating well-formed formulas, or WFFs.

An expression is a WFF if and only if one of the following three rules applies:

1. It is the single letter 'p', 'q', 'r', or 's'.
2. It is the letter 'N' followed by a legal WFF.
3. It is one of the letters 'C', 'A, 'K', or 'E' followed by two legal WFFs.

For example, the strings

p	[Rule 1]
Np	[Rule 2]
ApNp	[Rule 3]

are all WFFs as shown by applications of the indicated rule.

Each of the symbols that make up a WFF has an interpretation in propositional logic. The letters 'p', 'q', 'r,' and 's' represent propositional variables which ordinarily correspond to statements which are either true or false. Thus, 'p' might represent the statement "it is raining now," and 'q' the statement "I should carry an umbrella." The letters 'N', 'K', 'A', 'C', 'E' represent the logical functions "not," "and," "or," "if . . . then," and "is equivalent to," respectively. Thus, the WFF

Cpq

means "if it is raining now, then I should carry an umbrella."

Based on this idea:

(a)   Write a function **VALIDWFF** that reads in a line of characters from the terminal and returns **TRUE** or **FALSE** according to whether the line represents a valid **WFF**.

(b)   Design a data structure to represent **WFF**s internally.

(c)   Using much the same logic as in **VALIDWFF**, write a function **PARSEWFF** which returns the structure of the **WFF** it reads.

(d)   Write a function **EVALWFF** which takes the internal structure for a **WFF** and returns **TRUE** or **FALSE** depending on the truth value of the entire **WFF**, given the above interpretation of the connectives 'N', 'C', 'A', 'K', and 'E'. Assume that the variables **P, Q, R,** and **S** are defined as Boolean variables and that each contains the current value of the appropriate lowercase symbol.

Implementation of
Recursion

Never trust to general impressions, my boy, but concentrate yourself
upon details.
　　　　—Sherlock Holmes, A Case of Identity, *Sir Arthur Conan Doyle*

Up to this point, our concern has been with developing a strategy for recursive
programming based on a relatively general understanding of the underlying
mechanism. For the person who shares with Sherlock Holmes a desire to reduce
a problem to its smallest details, it is important to consider the implementation
of recursion from a more intimate perspective.

In Chapter 5, we developed a model for the implementation of recursion
using a stack of index cards to manage the necessary bookkeeping operations.
When faced with a recursive implementation, a computer system must keep
track of the same information in its internal storage. More specifically, for each
call to a procedure, we must (1) remember the values of all local variables and
(2) find some way to keep track of our progress through each of the subtasks.

The only thing that makes this at all difficult is the fact that we must
simultaneously maintain this information for each level in the recursive decomposition. Whenever a new procedure is called, a new environment is introduced which temporarily supersedes the previous one. When that procedure
returns, the old environment is restored.

11-1　The Control Stack Model

In order to accomplish this operation, most computers make use of special
hardware which implements a structure called the *control stack*. Internally,
the control stack consists of an array of consecutive memory locations associated with a stack pointer register (called **SP** in the subsequent text) which
points to the last item entered on the stack.

Manipulation of the stack is performed by two hardware operations traditionally called **PUSH** and **POP.** The **PUSH** operation takes a value and
stores it on the stack. For example, if we start with an empty stack, and push
the value 3, we are left in the position

Pushing the value 5 results in the configuration

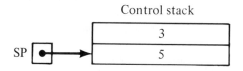

The **POP** operation returns the most recent value pushed onto the stack (in this case, the 5) and adjusts the **SP** register to indicate the next most recent value. Conventionally, the value 5 is not actually removed from storage, but it can no longer be accessed by the stack instructions:

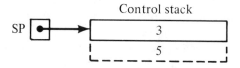

In a stack, the last value pushed is always the first value to be popped. For this reason, a stack is sometimes referred to as a LIFO storage mechanism, where LIFO is an acronym for "last in/first out." Significantly, this is precisely the behavior of procedures. The last procedure called is always the first from which we return. This similarity between the behavior of procedures and stack operation is the primary reason that stacks are chosen as the underlying data structure for keeping track of the control history.

Whenever a call is made to a procedure, the first step in the operation is to evaluate each of the arguments and place them on the stack. Thus, when the main program makes the call

MOVETOWER(3, 'A', 'B', 'C')

the stack will contain the values

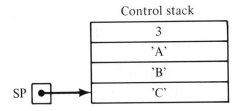

Control stack

The next step consists of transferring control to the procedure itself. Internally, this is implemented in much the same way as the **GOTO** statement, but with an additional hitch. When we make a procedure call, we must be able to return to exactly that point at which we left off. Thus, in the main program

> **BEGIN**
>   **WRITELN('Tower of Hanoi program for 3 disks.');**
>   **MOVETOWER(3, 'A', 'B', 'C');**
> →   **WRITELN('Finished.')**
> **END.**

we must be able to return to the point indicated by the arrow. This position is called the *return address* and is pushed on the stack as part of the call operation.

In a conventional machine, the return address will be the location number of the appropriate instruction in the stored program. At the Pascal level, however, we are never in a position to work with physical addresses, and it is easier to represent this in a symbolic way. Here, we will use the symbol <MAIN> to represent the address of the statement

> **WRITELN('Finished.')**

Thus, after the mechanics of executing the first call are complete, we are ready to begin the execution of **MOVETOWER** with the following values on the stack:

Control stack

3	N
'A'	START
'B'	FINISH
'C'	TEMP
< MAIN >	Return address

These five locations correspond to the first index card in the analysis given in Chapter 5. From the implementation point of view, the region of the stack which contains the data pertaining to each environment is called a *stack frame,* or *activation record.* As long as this environment is active, any references to the variables **N, START, FINISH,** or **TEMP** refer to the contents of the stack location with the corresponding label. Thus, when the program reaches the line

IF N = 1 THEN

the value of **N** is determined by checking the current stack frame. Since that value is currently 3, the **IF** condition is false, and the operation continues with the **ELSE** block of the **MOVETOWER** procedure.

Within that block, the first instruction encountered is

MOVETOWER(N–1, START, TEMP, FINISH)

which constitutes a recursive call to **MOVETOWER.** To make this call, we simply repeat the original process. First, the arguments are evaluated using the values in the *current* stack frame. Here, **N** is 3, **START** is 'A', **FINISH** is 'B', and **TEMP** is 'C', so that the above call is equivalent to

MOVETOWER(2, 'A', 'C', 'B')

Each of these values is pushed on the stack, followed by the return address. Once again, we will refer to these addresses symbolically by using the following labels:

```
         BEGIN
             MOVETOWER(N–1, START, TEMP, FINISH);
  <1> →      WRITELN(START, ' to ', FINISH);
             MOVETOWER(N–1, TEMP, FINISH, START);
  <2> →  END
```

Here, we are returning to label <1>, so that the new stack frame has the values:

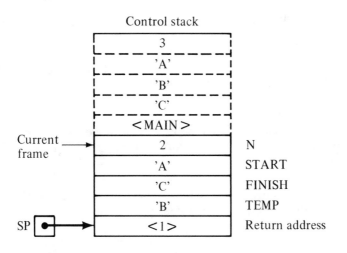

We can continue this analysis to any required depth, just as we did in the case of the index-card model. Thus, the next operation will be yet another call to **MOVETOWER** which generates a third stack frame:

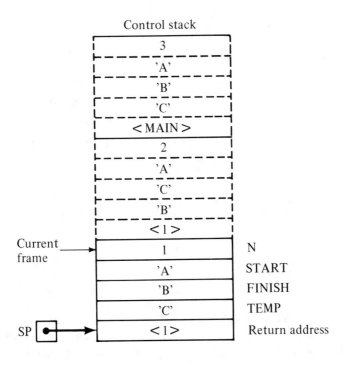

In this case, N is equal to 1, and we have therefore reached a simple case from which no additional recursive calls are required. The **WRITELN** statement is executed to produce the line

A to B

and the **MOVETOWER** procedure is ready to return.

Returning from a procedure consists of two steps:

1. Moving the current frame pointer back to indicate the previous frame
2. Transferring control to the address in the "return address" field on the stack

The return address here is <1>, so that we begin at label 1 in the **MOVETOWER** body after restoring the previous frame:

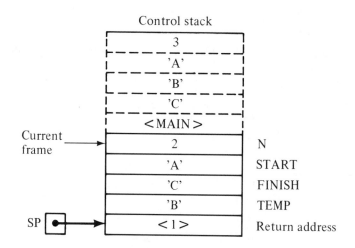

Control stack

From here, we simply continue the evaluation of **MOVETOWER**, performing call and return operations as indicated, until control returns to the main program.

11-2   Simulating Recursion

When using a language like Pascal, the stack operations described in the previous section are carried out automatically, and the programmer ordinarily remains oblivious to the details. In part, the attractiveness of recursion lies in the fact that much of this complexity is hidden away and the programmer can properly concentrate on the algorithm itself.

On the other hand, it is possible for the programmer to perform the stack operations explicitly and thereby *simulate* the recursive operation. When using a language in which recursion is not supported (such as FORTRAN, for example), defining an explicit control stack is often the only way to code a recursive algorithm. In Pascal, such an approach is less important, but it helps to provide a more intimate understanding of the underlying mechanics.

Following the lead of the hardware model described in the previous section, we will represent the stack using an array and a special stack pointer variable **SP**.

```
CONST
  MAXSTACK = 100;

VAR
  STACK : ARRAY [1..MAXSTACK] OF INTEGER;
  SP : INTEGER;
```

Here, we have designed the stack to contain integer values. This corresponds closely to the unit of a *memory word* in the underlying machine. To store other information (such as characters) in the stack, we will convert those values to their integer equivalents.

The **SP** variable keeps track of the number of items currently in the stack. Initially, the **SP** is therefore set to 0. The two principal stack operations are defined as follows:

```
PROCEDURE PUSH(X : INTEGER);
  BEGIN
    SP := SP + 1;
    STACK[SP] := X
  END;

PROCEDURE POP(VAR X : INTEGER);
  BEGIN
    X := STACK[SP];
    SP := SP - 1
  END;
```

To illustrate how the stack structure can be used to eliminate recursion from the Tower of Hanoi program, it helps to begin at the main program level. In our earlier implementation, the main program consisted of a simple call to **MOVETOWER** indicating the initial arguments:

$$MOVETOWER(3, 'A', 'B', 'C')$$

In the new version, **MOVETOWER** will no longer be able to use Pascal's conventional parameter passing mechanism and will expect its arguments on the control stack. Thus, the main program must store these arguments in the

correct locations in the stack frame. This can be done with four calls to the
PUSH procedure:

PUSH(3);
PUSH(ORD('A'));
PUSH(ORD('B'));
PUSH(ORD('C'));

For the last three values, it is necessary to use the **ORD** function to obtain the
ASCII values of the characters so that they can be stored as integers in the
stack.

On the other hand, writing four **PUSH** instructions each time we make a
new call to **MOVETOWER** seems somewhat tedious. Inside **MOVETOWER**,
we will also have occasion to push these arguments, and it will simplify these
programs if we define the following routine:

```
PROCEDURE PUSHARGS(N : INTEGER;
                   START,FINISH,TEMP : CHAR);
   BEGIN
     PUSH(N);
     PUSH(ORD(START));
     PUSH(ORD(FINISH));
     PUSH(ORD(TEMP));
   END;
```

This procedure handles the pushing of all four arguments simultaneously so
that we can think of this as a single operation.

In addition to pushing the arguments, the main program must complete
the first stack frame by pushing the appropriate return address. As noted above,
return addresses cannot be supplied explicitly in Pascal, but we can assign
numeric codes to refer to the possible return points. For consistency with the
previous section, we will define a constant **MAIN** and use that as a *sentinel* to
indicate a return to the main program. The actual value of **MAIN** is immaterial
(-1 was used here) but must not interfere with any other labels that we assign
during the implementation of **MOVETOWER**.

Thus, as it now stands, the main program consists of the following statements:

```
BEGIN
  SP := 0;
  PUSHARGS(3, 'A', 'B', 'C');
  PUSH(MAIN);
  MOVETOWER
END.
```

and produces the following stack frame:

STACK [1]	3	
STACK [2]	'A'	
STACK [3]	'B'	
STACK [4]	'C'	SP
STACK [5]	< MAIN >	5

From here, we can turn our attention to the **MOVETOWER** procedure itself. From this perspective, the most important thing to remember is that the parameters **N, START, FINISH,** and **TEMP** are no longer stored in simple variables but occupy specific positions within the stack frame as shown in the diagram below:

STACK[SP−4]	3	N
STACK[SP−3]	'A'	START
STACK[SP−2]	'B'	FINISH
STACK[SP−1]	'C'	TEMP
STACK[SP]	< MAIN >	Return address

Thus, every time we need to refer to the value of **N,** we must determine that value by finding the value in

$$STACK[SP-4]$$

Since this calculation is performed relative to the current setting of the stack pointer, this expression will always give the value of **N** in the currently active stack frame, as required by the recursive implementation.

To simplify references to arguments within the **MOVETOWER** routine, it is useful to define the names **N, START, FINISH,** and **TEMP** not as variables but as *functions* which access the correct location in the stack frame:

```
FUNCTION N : INTEGER;
  BEGIN
    N :=  STACK[SP−4]
  END;

FUNCTION START : CHAR;
  BEGIN
    START :=  CHR(STACK[SP − 3])
  END;
```

```
FUNCTION FINISH : CHAR;
  BEGIN
    FINISH := CHR(STACK[SP – 2])
  END;

FUNCTION TEMP : CHAR;
  BEGIN
    TEMP := CHR(STACK[SP – 1])
  END;
```

These functions also handle the conversion of the integer stack values back into their character equivalents. Thus, it is now possible to write the statement

WRITELN(START, ' to ', FINISH);

instead of the more cryptic

WRITELN(CHR(STACK[SP-3]), ' to ', CHR(STACK[SP-3]));

The only remaining task is the elimination of the recursive calls inside **MOVETOWER**. In the recursive version, the body of **MOVETOWER** appears as follows:

```
IF N = 1 THEN
  WRITELN(START, ' to ', FINISH)
ELSE
  BEGIN
    MOVETOWER(N – 1, START, TEMP, FINISH);
    WRITELN(START, ' to ', FINISH);
    MOVETOWER(N – 1, TEMP, FINISH, START)
  END
```

and contains two recursive calls. For each of these, the revised code must perform the following operations:

1. Using the **PUSHARGS** routine defined above, each of the arguments in the call must be pushed onto the stack to begin the definition of the next frame.*
2. A value must be pushed on the stack to indicate the return address for this call. Here the two possible return points are given the labels **1:** and **2:** (after the first and second calls, respectively) and the return address is indicated by pushing the corresponding integer.

---

*Here, the use of **PUSHARGS** turns out to be quite important. If this operation were performed using four separate calls to **PUSH**, the stack pointer would change between each operation which, in turn, would affect the calculation of the arguments. Here, all the arguments are evaluated using the old value of **SP** and passed to **PUSHARGS** where the stack pointer is actually changed.

3. The program must jump back to the first instruction in the **MOVE-TOWER** procedure. Here this is accomplished by labeling the first statement as **0:** and using the statement **GOTO 0.**

Thus, up to this point, the new body of **MOVETOWER** appears as follows:

```
0:  IF N = 1 THEN
        WRITELN(START, ' to ', FINISH)
    ELSE
      BEGIN
        PUSHARGS(N - 1, START, TEMP, FINISH);
        PUSH(1);
        GOTO 0;
1:      WRITELN(START, ' to ', FINISH);
        PUSHARGS(N - 1, TEMP, FINISH, START);
        PUSH(2);
        GOTO 0;
2:      ;
      END;
```

It is important to recognize how closely this program corresponds to the original. The only transformations that have been made are replacing any recursive call of the form

MOVETOWER(n, start, finish, temp);

with the lines:

```
            PUSHARGS(n, start, finish, temp);
            PUSH(label);
            GOTO 0;
   label:
```

where "label" is some integer label uniquely identifying the return point.

To finish this program, we must simulate the return from the **MOVE-TOWER** procedure. Returning from **MOVETOWER** requires clearing the previous stack frame and jumping to the label indicated by the return address. This is accomplished by the following code:

```
            POP(RETADD);
            FLUSH(4);
            CASE RETADD OF
                1 : GOTO 1;
                2 : GOTO 2;
              MAIN : (* return to main program *);
            END;
```

The procedure **FLUSH(N)** simply pops the top **N** elements from the stack and throws them away. This can be accomplished merely by adjusting the stack pointer as shown in the following code:

```
PROCEDURE FLUSH(N : INTEGER);
BEGIN
    SP := SP-N
END;
```

The **POP** call takes the return address marker off the control stack and stores it in the variable **RETADD**. The **FLUSH(4)** statement pops off the four parameters, restoring the previous frame. Finally, the **CASE** statement interprets the value of **RETADD** and jumps to the appropriate label. It is important to note that labels and integers are not at all the same object, and it is impossible to write **GOTO RETADD** as a statement in Pascal. You should also notice that if **RETADD** has the value **MAIN** (− 1), the **CASE** statement performs no action and **MOVETOWER** returns to the main program.

Thus, the complete nonrecursive coding for **MOVETOWER** is:

```
PROCEDURE MOVETOWER;
VAR
    RETADD : INTEGER;
BEGIN
0:  IF N = 1 THEN
        WRITELN(START, ' to ', FINISH)
    ELSE
      BEGIN
        PUSHARGS(N−1, START, TEMP, FINISH);
        PUSH(1);
        GOTO 0;
1:      WRITELN(START, ' to ', FINISH);
        PUSHARGS(N−1, START, TEMP, FINISH);
        PUSH(2);
        GOTO 0;
2:      ;
      END;
    POP(RETADD);
    FLUSH(4);
    CASE RETADD OF
        1 : GOTO 1;
        2 : GOTO 2;
      MAIN : (* return to main program *);
    END;
END;
```

Go through this program carefully and make sure that you can follow its internal operation.

At the same time, it is important to keep in mind that this sort of implementation is not required if the language permits recursive procedure calls. Compared with the recursive version presented in Chapter 5, this new implementation is both less clear and less efficient.

Bibliographic Notes

To understand the material in this chapter, it is most useful to work in the context of a specific computer architecture and to examine the operation of recursive algorithms in that environment. Thus, some of the most useful texts in this regard are assembly-language handbooks for the computer system you use. However, reasonably good machine-independent discussions of the implementation of recursion can be found in Barron [1968] and Tenenbaum and Augenstein [1981].

Bibliography

This bibliography is by no means complete, but it does provide several useful sources for further reading in addition to those listed in the text.

Abelson, Harold, and Andrea diSessa, *Turtle Geometry,* MIT Press, Cambridge, Mass., 1981.

Aho, Alfred V., and Jeffrey D. Ullman, *Principles of Compiler Design,* Addison-Wesley, Reading, Mass., 1979.

————, John Hopcroft, and Jeffrey Ullman, *The Design and Analysis of Algorithms,* Addison-Wesley, Reading, Mass., 1974.

Augenstein, Moshe J., and Aaron M. Tenenbaum, "A Lesson in Recursion and Structured Programming," *SIGCSE Bulletin,* vol. 8, no. 1, February 1976.

Barron, D. W., *Recursive Techniques in Programming,* American-Elsevier, New York, 1968.

Bentley, Jon, "Programming Pearls," *Communications of the ACM,* vol. 27, no. 4, April 1984.

Berliner, Hans J., "A Chronology of Computer Chess and Its Literature," *Artificial Intelligence,* vol. 10, no. 2, 1978.

Bird, R. S., "Improving Programs by the Introduction of Recursion," *Communications of the ACM,* vol. 20, no. 11, November 1977.

————, "Notes on Recursion Elimination," *Communications of the ACM,* vol. 20, no. 6, June 1977.

Cherniak, E., C. Riesbeck, and D. McDermott, *Artificial Intelligence Programming,* Lawrence Erlbaum Associates, Hillsdale, N.J.

Cohen, Joel, "On the Nature of Mathematical Proofs," *The Worm-Runner's Digest,* vol. 3, no. 3, December 1961.

Cooper, Doug, and Paul Clancy, *Oh! Pascal!,* W. W. Norton, New York, 1982.

Dewdney, A. K., "Computer Recreations—Tower of Hanoi," *Scientific American*, August 1983.

———, "Computer Recreations," *Scientific American*, July 1984.

Foley, J., and A. Van Dam, *Fundamentals of Interactive Computer Graphics*, Addison-Wesley, Reading, Mass., 1982.

Ford, Gary A., "A Framework for Teaching Recursion," *SIGCSE Bulletin*, vol. 14, no. 2, June 1982.

———, "An Implementation-Independent Approach to Teaching Recursion," *SIGCSE Bulletin*, vol. 16, no. 1, February 1984.

Grogono, Peter, *Programming in Pascal*, 2d ed., Addison-Wesley, Reading, Mass., 1984.

Hofstadter, Douglas, "Metamagical Themas—LISP," *Scientific American*, March 1983.

———, *Godel, Escher, Bach: An Eternal Golden Braid*, Basic Books, New York, 1979.

Jensen, Kathleen, and Niklaus Wirth, *Pascal User Manual and Report*, 2d ed., Springer-Verlag, New York, 1971.

Knuth, Donald, *The Art of Computer Programming—Fundamental Algorithms*, 2nd ed., Addison-Wesley, Reading, Mass. 1973.

———, *The Art of Computer Programming—Sorting and Searching*, Addison-Wesley, Reading, Mass., 1973.

Levitt, Ruth, ed., *Artist and Computer*, Harmony Books, New York, 1976.

Papert, Seymour, *Mindstorms*, Basic Books, New York, 1981.

Polya, Gyorgy, *How to Solve It*, Doubleday, Garden City, N.Y., 1957.

Reingold, Edward M., and Wilfred J. Hansen, *Data Structures*, Little, Brown and Company, Boston, 1983.

Sedgewick, Robert, *Algorithms*, Addison-Wesley, Reading, Mass., 1983.

Shannon, Claude, "Automatic Chess Player," *Scientific American*, vol. 182, no. 48, 1950.

Solow, Daniel, *How to Read and Do Proofs*, John Wiley, New York, 1982.

Tenenbaum, Aaron M., and Moshe J. Augenstein, *Data Structures Using Pascal*, Prentice-Hall, Englewood Cliffs, N.J., 1981. Weizenbaum, Joseph, *Computer Power and Human Reason*, W. H. Freeman, New York, 1976.

Wickelgren, Wayne A., *How to Solve Problems*, W. H. Freeman, New York, 1974.

Winston, Patrick, and Berthold K. P. Horn, *LISP*, Addison-Wesley, Reading, Mass., 1981.

———, *Artificial Intelligence*, 2nd ed., Addison-Wesley, Reading, Mass., 1984.

Wirth, Niklaus, *Algorithms + Data Structure = Programs*, Prentice-Hall, Englewood Cliffs, N.J., 1976.

Index